GW00502591
This book belongs to...

Contents

Packed with Raspberry Pi fun and games!

Feature

Getting Started...

Bringing your Raspberry Pi to life!

Got a Raspberry Pi? Great! There's a wonderful world of computing fun waiting for you. Let's get started by making sure you have all the cables and accessories you will need, and showing you how to plug them all in. We will soon have your Pi adventure up and running...

What you will learn

In this beginners' feature we will show you...

- How to select the right equipment for your Pi adventure
- How to connect everything together correctly before you start
- How to download and set up the software you will need

What you will need

We're going to explain each in detail, but here are the bits and pieces of tech you need...

- Raspberry Pi
- Monitor or TV
- HDMI cable
- USB keyboard
- USB mouse
- Power supply
- microSD card

6 The Raspberry Pi Annual 2018

Minecraft

Coding with Minecraft

Programming the game on the Pi!

Minecraft is a popular sandbox open-world building game. A free version of Minecraft is available for the Raspberry Pi; it also comes with a programming interface. This means you can write commands and scripts in Python code to build things in the game automatically. It's a great way to learn Python!

What you will learn

By following this project with your Raspberry Pi, you will learn...

- How to access Minecraft Pi and create a new world
- How to use the Python programming environment IDLE to connect to Minecraft Pi
- How to use the Minecraft Python API to post text to the chat window, find the player's coordinates, teleport and build structures
- How to use variables to store IDs for different types of blocks
- Experimenting with placing different types of blocks with special attributes

Basics

Check off all your kit, set up your Raspberry Pi, and install the software for fun!

THE ALMOST CERTAINLY TRUE STORY OF RASPBERRY PI

Comic

The evil Doctor Bluescreen is uploading all the kids. Can the Pi Crew save the day?

We've hidden 12 Raspberry Pi logos around the book. Can you spot them all?

Coding

Begin with Scratch and move on to Python with our fun Pi coding projects!

Starting page 14

Scratch Coding

Learn the Basics of... Scratch 2

Python Project

Turtley Amazing

Move from Scratch to Python with the Turtle!

In this project you will take your first steps with the programming language Python to draw shapes, patterns, and spirals. You will use a module named Turtle. Along the way you will learn how to think in sequences, and use loops to repeat a sequence. This is a great stepping stone from a visual programming language like Scratch to the text-based environment of Python.

What you will learn

What you will need

Note

Tick off each step as you go!

64 The Raspberry Pi Annual 2018

Puzzles

Test your brains with our Pi Wordsearch, Spot the Differences, Maze and more!

Starting page 13

Puzzles

Space Maze

Take a trip to outer space!

Timed Challenge!

The Result!

46 The Raspberry Pi Annual 2018

Features

Getting Started...

Bringing your Raspberry Pi to life!

Got a Raspberry Pi? Great! There's a wonderful world of computing fun waiting for you. Let's get started by making sure you have all the cables and accessories you will need, and showing you how to plug them all in. We will soon have your Pi adventure up and running...

What you will learn

In this beginners' feature we will show you...

- How to select the right equipment for your Pi adventure
- How to connect everything together correctly before you start
- How to download and set up the software you will need

What you will need

We're going to explain each in detail, but here are the bits and pieces of tech you need...

- Raspberry Pi
- Monitor or TV
- HDMI cable
- USB keyboard
- USB mouse
- Power supply
- microSD card

1. The Raspberry Pi

The Raspberry Pi 3 is the third version of Raspberry Pi. It replaced the Raspberry Pi 2 Model B in February 2016. New features include...

- A 1.2GHz 64-bit quad-core ARMv8 CPU
- 802.11n Wireless LAN
- Bluetooth 4.1
- Bluetooth Low Energy (BLE)

Like the Pi 2, it also has...

- 4 USB ports
- 40 GPIO pins
- Full HDMI port
- Ethernet port
- Combined 3.5 mm audio jack and composite video
- Camera interface (CSI)
- Display interface (DSI)
- microSD card slot (now push-pull rather than push-push)
- VideoCore IV 3D graphics core

The Raspberry Pi 3 looks the same as the previous Pi 2 (and Pi 1 Model B+) and has complete compatibility with Raspberry Pi 1 and 2.

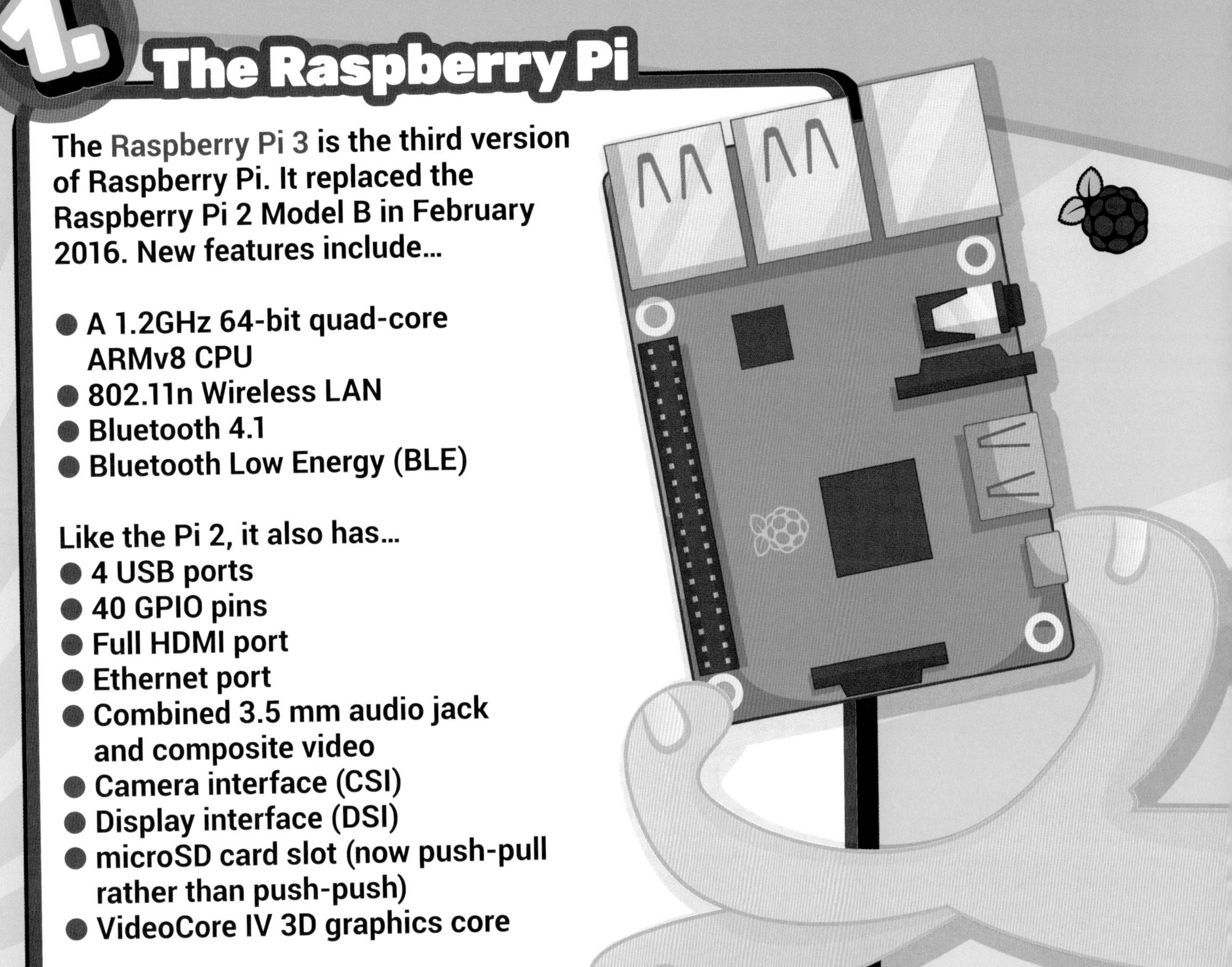

2. Monitor or TV

A monitor or TV with HDMI in can be used as a display with a Raspberry Pi. This is the quickest and easiest way to see what your Pi is thinking. How big a screen you want is up to you!

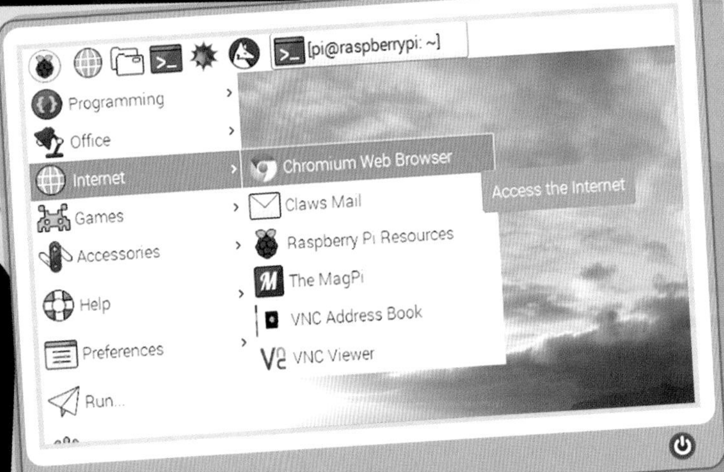

3. HDMI to HDMI Cable

Connect your Raspberry Pi to a monitor or TV with an HDMI cable. It'll give you great-quality pictures.

4. USB Keyboard

A USB keyboard is used to input text into a computer, laptop or a Raspberry Pi. Any keyboard will work, so you don't need a special one. Plug-and-play keyboards will work without any additional driver. Simply plug one into the Raspberry Pi and it should be recognised when it starts up.

5. USB Mouse

A USB mouse is used to move the cursor or mouse pointer around the screen. It's called a mouse as it looks like it has a tail! Like the keyboard, it will be automatically recognised by the Pi when plugged in.

6. Power Supply

If you're using a Raspberry Pi 3, then it's recommended that you use a 5 V, 2.5 A power supply. Earlier models can be powered using a current as low as 1.2 A. Ideally, you want to use a power supply which you know to be safe and which provides enough power to the Pi. You can buy the official Raspberry Pi power supply, or you can use a 5 V micro USB charger, the kind used by many mobile devices. You do need to check that your power supply provides the right voltage and current (5 V / 1.2–2.5 A).

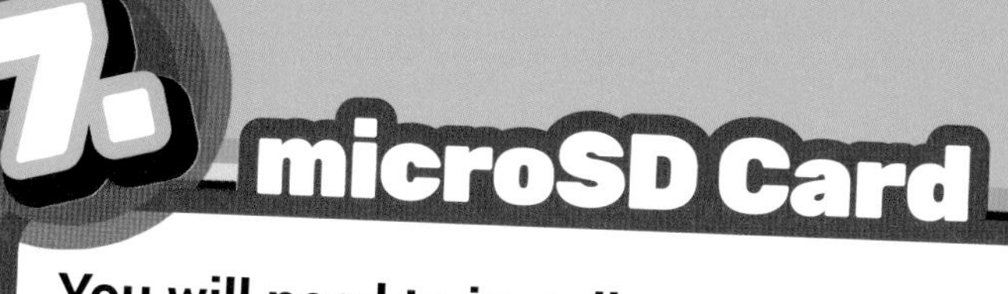

7. microSD Card

You will need to install an operating system on to your Pi so it knows what to do. The latest version of Raspbian, the default operating system recommended for the Raspberry Pi, needs an 8GB (or bigger) microSD card. Not all SD cards are the same, and some can fail more than others. If you're unsure, you can always buy the official Pi SD cards. Any 8GB SD card will work.

8. Plugging in your Raspberry Pi

1. Begin by placing your microSD card into the SD card slot on the Raspberry Pi. It will only fit one way, so you can't get it wrong.
2. Next, plug your keyboard and mouse into the USB ports on the Raspberry Pi. They will be automatically recognised.
3. Make sure that your monitor or TV is turned on, and that you have selected the right input (e.g. HDMI 1, HDMI 2, etc).
4. Connect your HDMI cable from your Raspberry Pi to your monitor or TV.
5. If you intend to connect your Raspberry Pi to the internet, plug an Ethernet cable into the Ethernet port, or connect a WiFi dongle to one of the USB ports (unless you have a Raspberry Pi 3, which doesn't need it).
6. When you're happy that you have plugged all the cables and SD card in correctly, connect the micro USB power supply. This action will turn on and boot your Raspberry Pi.

9. Connecting to the Internet

You will probably want to connect your Raspberry Pi to your local network or the internet. You can use any of the following options to do this...

● Connecting via Ethernet

The Raspberry Pi has an Ethernet port, alongside the USB ports. If your Pi is situated close to a router, access point, or switch, you can connect to a network using an Ethernet cable.

Once you've plugged the Ethernet cable into the Raspberry Pi and the other end into an access point, your Raspberry Pi will automatically connect to the network.

● Connecting via WiFi

If you have a Raspberry Pi 3, it has built-in wireless LAN. If you're using an earlier version of the Raspberry Pi, then you will need a USB WiFi dongle. Some WiFi dongles, when used with the Raspberry Pi, are simple plug-and-play devices. Others require specific drivers, and may not be compatible with the Raspberry Pi. Make sure you read the device manufacturer's documentation before buying one.

10. Adding Audio Output

What's the point in your Raspberry Pi making lots of great noises if you can't hear them?

● 3.5 mm audio port

The Raspberry Pi comes with a 3.5 mm audio port. This will allow you to plug most speakers or headphones into the Pi so that you can listen to the output from fantastic programs like Sonic Pi.

● Bluetooth speakers

With either the Raspberry Pi 3 or a Bluetooth dongle, you can connect to Bluetooth speakers or headphones. Your success rate may vary depending on the dongle and/or speakers you're using, so ensure that you read the manufacturer's documentation before you buy.

Adding Storage to your Raspberry Pi

You might find that the 8GB SD card you're using with your Raspberry Pi just isn't big enough for your needs. There are several options for increasing the storage capacity of your Raspberry Pi...

A bigger SD card

microSD cards come in a variety of sizes. The largest (reasonably priced) microSD cards are 128GB, which will provide you with lots of storage. As always, check the manufacturer's documentation to ensure that the card is compatible with a Raspberry Pi.

USB flash drives

There are lots of different types of storage device which you can plug directly into your Raspberry Pi's USB ports. USB flash drives come in a variety of styles and sizes, and can offer up to 1TB of space if you need that much.

External hard drives

You can also purchase external hard drives which can be connected via a USB cable. You have to be a little careful here. Some external hard drives are independently powered, and will work without problems. Some draw their power via the USB port, and might need more current that the Raspberry Pi can supply. Read the manufacturer's documentation to ensure any hard drive you're using will work with the Raspberry Pi. Some external hard drives have been designed specifically to work with the Raspberry Pi, such as the WD PiDrive 314GB.

11. Installing Raspbian on your SD Card with NOOBS

So you've just got hold of your first Raspberry Pi and you need to get the software up and running? Let's show you how to install Raspbian on your Raspberry Pi in no time at all...

Downloading NOOBS

Using NOOBS is the easiest way to install Raspbian on your SD card. To get hold of a copy of NOOBS...

1. Visit www.raspberrypi.org and click on the Downloads button in the navigation bar, at the top of the screen.
2. You should see a box which contains a link to the NOOBS files. Click on the link.
3. The simplest option is to download the zip archive of the files.

Writing NOOBS to an SD Card

Visit etcher.io and download and install the Etcher SD card image utility. Run Etcher and select the Raspbian zip file you downloaded. Select the SD card drive – Etcher will usually do this for you. Finally, click Burn to transfer NOOBS to the SD card. Once complete, the utility will eject/unmount the SD card so it's safe to remove it from the computer.

Booting from NOOBS

1. Once the files have been copied over, insert the microSD card into your Raspberry Pi and then plug it into a power source.
2. You will be provided with a single option, once the installer has loaded. You should check the box for Raspbian, and then click Install.
3. Click Yes at the warning dialog, and then sit back and relax. It will take a while, but Raspbian will install.

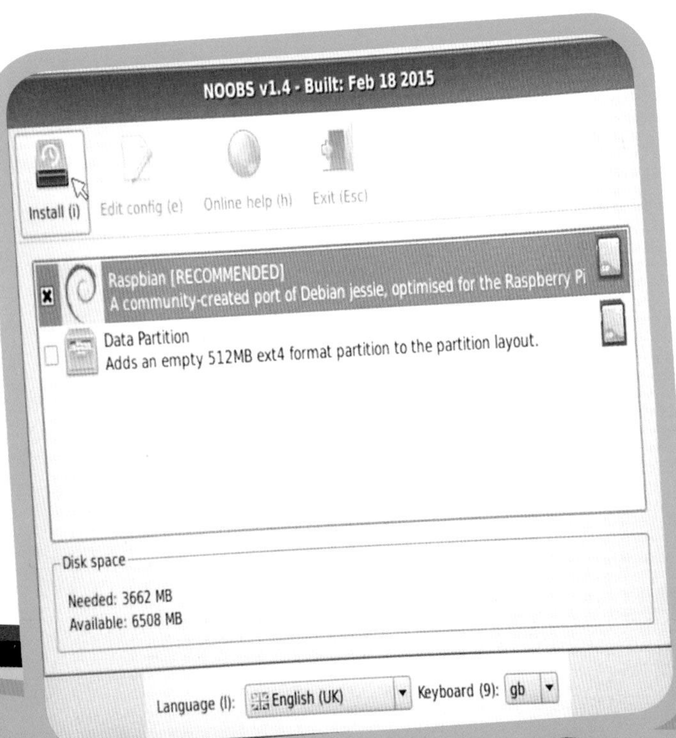

Mad Pi Dash

Who is plugged into the Pi?

Everyone wants a piece of **Raspberry Pi**! Here are five characters from the world of Pi. All of them want to plug in and create a cool project, but **which one is connected?** Trace the cables to find out!

Learn the Basics of... Scratch 2

Create stories, games and animations!

Tick off each step as you go!

Scratch is a visual programming tool which features a very easy-to-use drag-and-drop interface. It enables you to create your own computer games, interactive stories, and animations using some programming techniques without actually having to write code. This feature will help get you started with the basics of Scratch.

What you will learn

By following the steps in this feature, you will learn...

- What all the buttons and toolbars in the Scratch window do
- How to use blocks to make the Scratch cat move about
- How to change sprites
- How to create your own sprites

What you will need

This tutorial requires Scratch 2. To use it, you need a Raspberry Pi 2 or 3 running the latest version of the Raspbian operating system, plus a standard USB keyboard and mouse.

HEE-HEE!
I CAN'T WAIT
TO GET STARTED
IN SCRATCH!

1. Open up Scratch

You'll find Scratch 2 in Menu > Programming. Once opened, you will see a window like this…

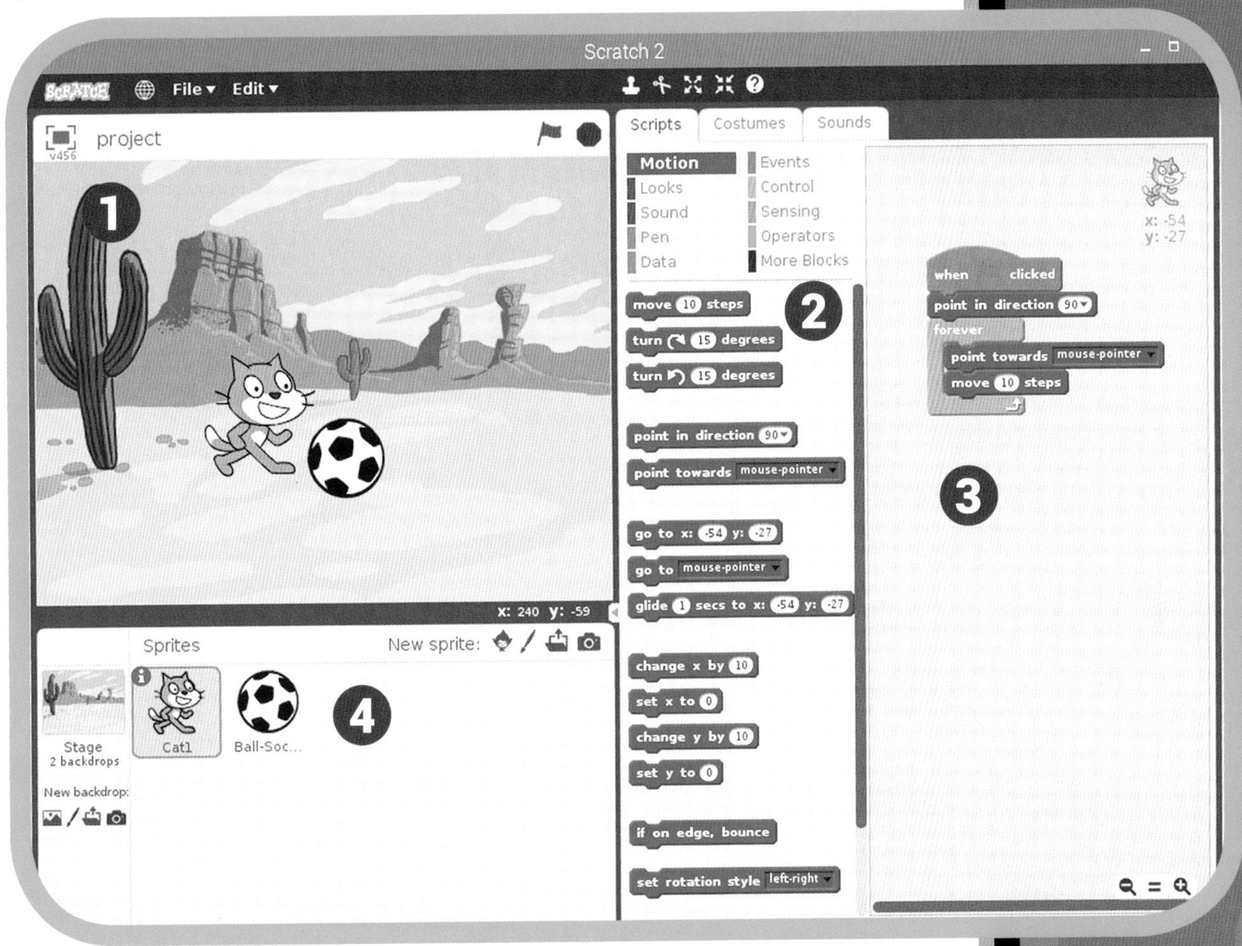

1. Stage
2. Blocks palette
3. Scripts area
4. Sprite list

2. Make the Scratch cat move

The cat on the white background is an example of a sprite in Scratch. Currently the Scratch cat sprite is on a blank stage. First, let's get the cat to move.

1. Click on the Scratch cat sprite.

2. Then click on the blocks palette and select Events.

3. Next, drag a when green flag clicked block and place it on the scripts area on the right of the screen.

4. Add a blue move 10 steps block from the Motion blocks palette and connect it to the when green flag clicked block.

5. Now click the green flag icon at the top right of the stage and see the cat move!

6. How would you get the Scratch cat to move further?

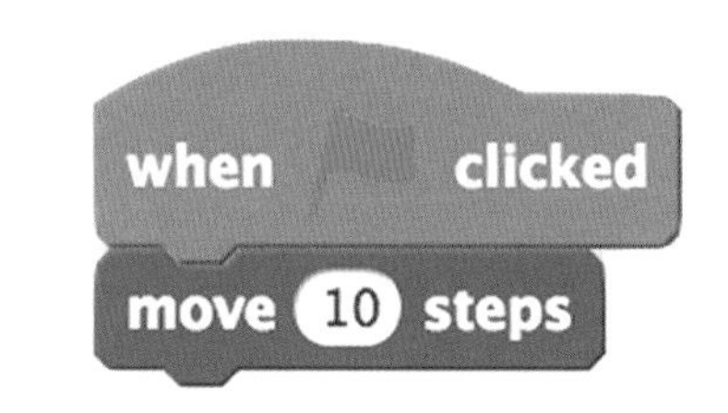

3. Changing the way the sprite looks

All sprites in Scratch can wear different costumes.

1. Click on your sprite to select it. In the Scripts area in the middle of the screen, click on the Costumes tab.

2. You will see that the cat has two costumes. Right-click costume 2 and select duplicate to make a third costume.

3. Select costume3 and it will appear in the Paint Editor. Experiment with all the buttons and tools to find out what they do.

4. Next, draw some clothes on the costume and click OK.

5. To switch between costumes, click on the Scripts tag. Add the purple Looks block, switch to costume, to the other blocks and connect it.

6. Select costume3 from the drop-down menu on the purple block.

7. Now run your program to see the costume change.

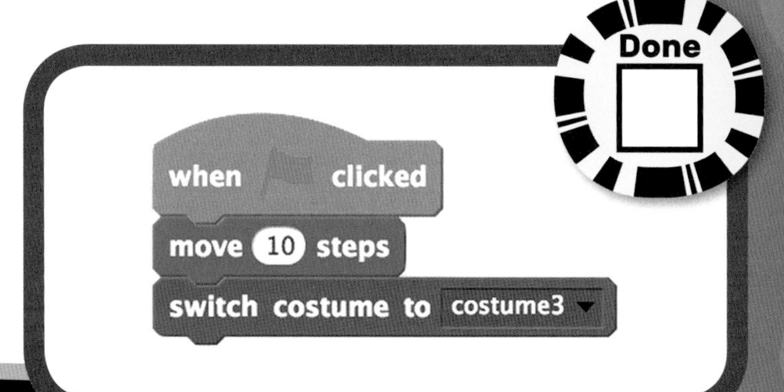

4. Add sprites

If you feel that the Scratch cat sprite does not suit your needs then you can add other sprites, or even create your own!

1. At the top of the Sprites palette are four icons to create a new sprite.
2. The first allows you to Choose sprite from library. This opens a window where you can choose one of Scratch's built-in sprites.
3. The second icon, Paint new sprite, opens the Paint Editor. Here you can use shapes, lines, and freehand drawings to make your own custom characters. Have fun!
4. The third icon lets you Upload sprite from file, to use an existing image or Scratch sprite.
5. The fourth icon enables you to create a New sprite from camera (if one is connected).

Sprites New sprite:

Sprite1

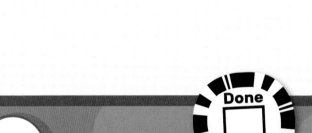

SCRATCH IS A GREAT INTRODUCTION TO CODING!

What next?

Now you know the basics of Scratch, there are lots of great projects you can have a go at...

- Try completing the Robot Antenna Resource to make an LED blink!
- Create a button using candy sweets with the Sweet Shop Reaction Game.

These and more can be found on the Raspberry Pi Learning Resources website at...
www.raspberrypi.org

Physical Computing With Scratch

Turn simple code into real life!

What you will learn

By completing this project you will learn...

- How to control the GPIO pins using Scratch
- How to receive input from the GPIO pins using Scratch

The version of Scratch included with the Raspberry Pi has a number of unique features; one of the most useful is its ability to communicate with the GPIO pins (General Purpose Input Output). These pins allow you to connect your Raspberry Pi to a range of devices, from lights and motors to buttons and sensors. The original Raspberry Pi had a 26-pin header and newer models (B+, Pi 2, Pi 3) have a 40-pin header, but this project will work with any model.

Note

Visit the Raspberry Pi website for latest stockist information for the extra kit you will need to do this Scratch project.
www.raspberrypi.org

What you might need

As well as a Raspberry Pi with an SD card and the usual peripherals, you'll also need...

- MALE-TO-FEMALE JUMPER CABLE
- TACTILE PUSH BUTTON
- BREADBOARD
- PIR SENSOR
- 330R RESISTOR
- LED
- PIEZO BUZZER

Tick off each step as you go!

COMBINE ME WITH SCRATCH AND WE CAN DO GREAT THINGS TOGETHER!

1. GPIO pins

One powerful feature of the Raspberry Pi is the row of GPIO pins along the top edge of the board. GPIO stands for General-Purpose Input/Output. These pins are a physical interface between the Raspberry Pi and the outside world. At the simplest level, you can think of them as switches that you can turn on or off (input) or that the Pi can turn on or off (output).

The GPIO pins allow the Raspberry Pi to control and monitor the outside world by being connected to electronic circuits. The Pi is able to control LEDs, turning them on or off, run motors, and many other things. It's also able to detect whether a switch has been pressed, the temperature, and light. We refer to this as physical computing.

There are 40 pins on the Raspberry Pi (26 pins on early models), and they provide various different functions.

If you have a RasPiO pin label, it can help to identify what each pin is used for. Make sure your pin label is placed with the keyring hole facing the USB ports, pointed outwards.

If you don't have a pin label, then this guide can help you to identify the pin numbers...

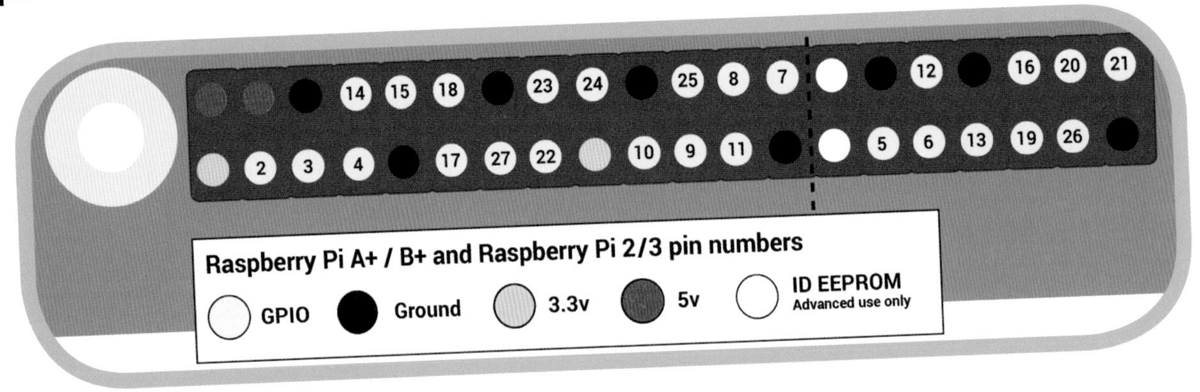

You'll see pins labelled as 3V3, 5V, GND and GP2, GP3, etc...

3V3	3.3 volts	Anything connected to these pins will always get 3.3 V of power
5V	5 volts	Anything connected to these pins will always get 5 V of power
GND	ground	Zero volts, used to complete a circuit
GP2...	GPIO pin 2...	These pins are for general-purpose use and can be configured as input or output pins
ID_SC/ID_SD/DNC	Special purpose pins	Don't use these ones

Done

WARNING!

You must be careful with the pins on your Raspberry Pi or you can do it irrepairable damage.

If you follow the instructions, then playing about with the GPIO pins is safe and fun. Randomly plugging wires and power sources into your Pi, however, may destroy it, especially if using the 5V pins. Bad things can also happen if you try to connect things to your Pi that use a lot of power; LEDs are fine, motors are not. If you're worried about this, then you might want to consider using an add-on board such as the Explorer HAT until you're confident enough to use the GPIO directly.

2. Lighting an LED

You can test whether your GPIO pins and LEDs are working by building the circuit below. You can use any resistor over about 50Ω.

1. The LED is connected directly to the GND pin, and the 3V3 pin via the resistor, and should light up.

2. Be sure to connect your LED the correct way round; the longer leg should be connected to the 3V3 pin...

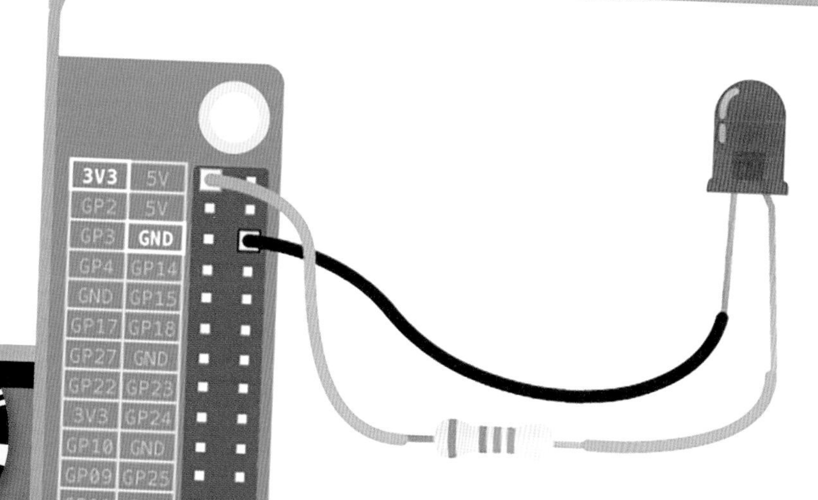

Done

3. Using a switchable pin

1. To control the LED, you'll need to adapt your circuit to use a switchable pin.

2. In the diagram here, pin 17 has been used, but you can use any numbered pin you wish.

4. Constructing a Scratch program

1. Locate the Scratch program by clicking on Menu followed by Programming, and selecting Scratch 2.

2. The familiar Scratch interface will then load...

3. Click on Events from the blocks palette. Drag the when green flag clicked block onto the scripts area...

when clicked

4. To add GPIO functionality, first click More Blocks and then Add an Extension. You should then select the Pi GPIO extension option and click OK.

5. In the More Blocks section, you'll now see two additional blocks for controlling and responding to your Pi GPIO pins. You can change the pin number by selecting the round field and typing it in.

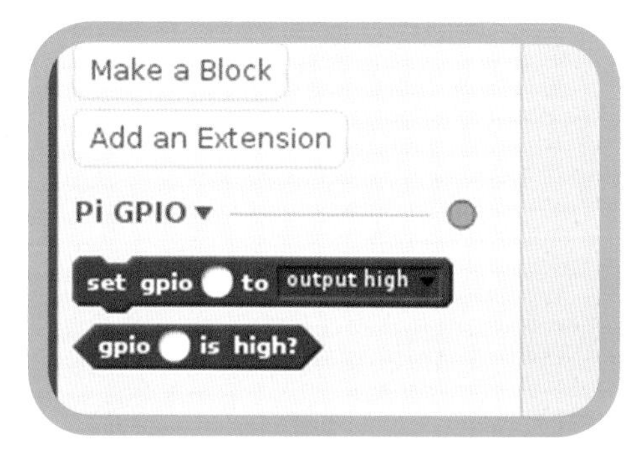

6. With set gpio 17 to output high or low, you can turn on your LED attached to GPIO pin 17 on or off. Using two of these inside a forever block, with wait blocks to add a delay, you can make the LED flash continuously...

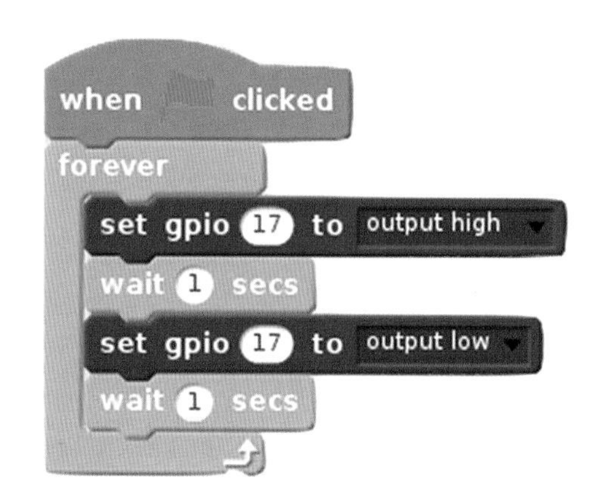

5. Connecting a button

1. As well as controlling the physical world, you can react to it using an input device such as a button.

2. Connect your button to a breadboard, then connect one pin to a ground pin and the other to a numbered GPIO pin. In this example pin 2 has been used...

6. Configuring your button

1. Before Scratch can react to your button, it needs to be told which GPIO pin is configured as an input pin.

2. If you have added the Pi GPIO extension (step 4, page 22), clicking More Blocks will show a couple of blocks related to the Pi's GPIO pins.

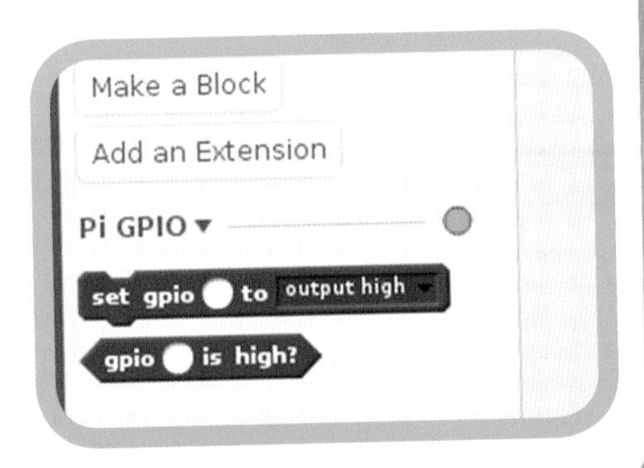

3. In a new Scratch project file (File > New), select Events from the blocks palette and add a when green flag clicked to the scripts area.

4. Select More Blocks, then drag a set gpio to block under the green flag one.

5. Alter the block's number field to 2 – to set it to GPIO 2 for your button – and use its drop-down menu to set it to an input.

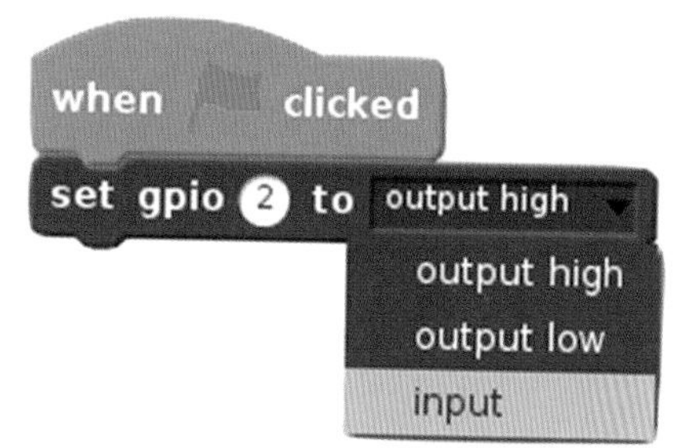

6. Clicking the green flag will now set GPIO pin 2 to an input, so we will be able to sense whether the button is being pressed.

7. Pressing the button right now won't do anything, but we can create a fairly simple program to respond to a button press to trigger something...

Done

7. Responding to a button press

1. Now that your button is all set up and working, you can make it do something. You can start off by making it control a sprite.

2. Begin with a forever loop with an if else block inside it. This will continually check the if condition and perform some action if the condition is met or not. In this case showing one of two messages.

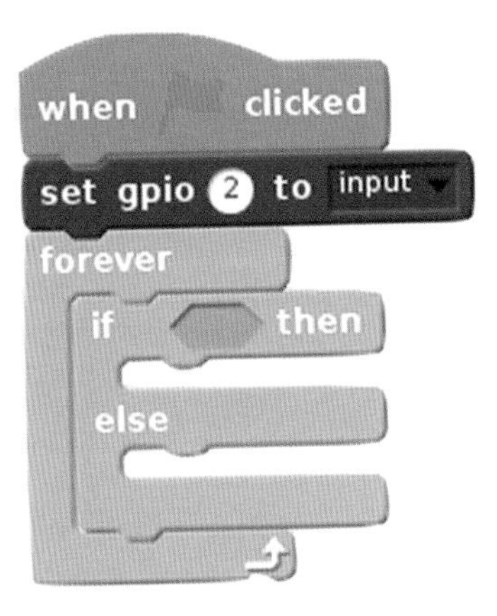

3. Now you need to add the condition, which requires a gpio 2 is high? block to be placed in the if else block's field. Note that as the pin is set high by default, and the button pulls it low, we put the say Hello block under else.

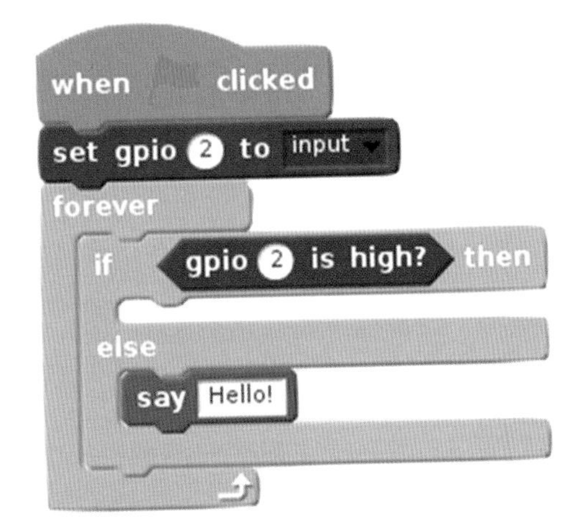

Done

If everything is correct, your button should make the sprite say 'Hello'.

8. Controlling an LED with a button push

To finish off, you can combine your two programs so that the button can turn the LED on and off.

1. Adapt your script by replacing the say blocks in the if else block with set gpio 17 to output low and high...

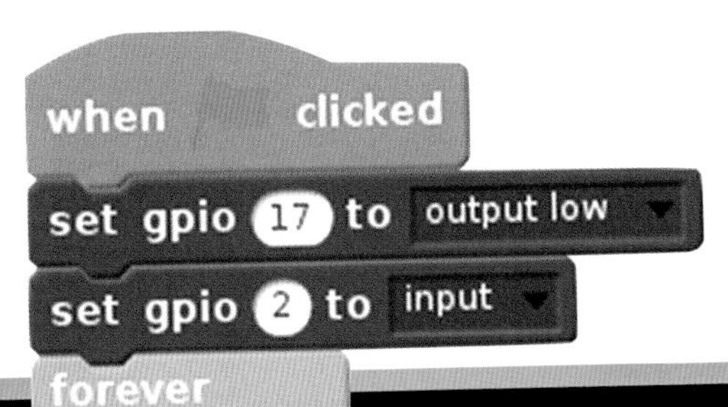

when clicked
set gpio 17 to output low
set gpio 2 to input

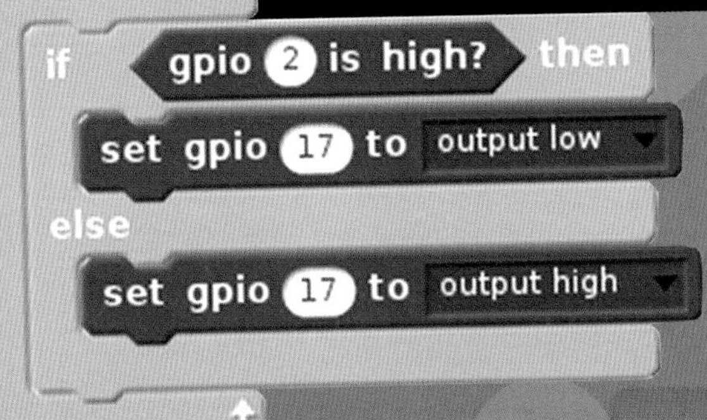

forever
if gpio 2 is high? then
set gpio 17 to output low
else
set gpio 17 to output high

2. Now when you push the button, the LED should light up.

Done

What next?

There are lots of other things you can control or monitor with your Raspberry Pi. Have a look at the worksheets on the website to see how easily this can be done...

- Using an active buzzer
- Making traffic lights
- Using a PIR sensor

These and more can be found on the Raspberry Pi Learning Resources website at... www.raspberrypi.org

Tick off each step as you go!

Santa Detector

See who you can catch on Christmas Eve!

HOW DARE YOU THINK YOU COULD EVER CATCH ME!

No more propping your eyes open with matchsticks to try to catch Santa! This Scratch program uses a low-cost, infrared detector to set off an alarm when Santa sneaks into your room. It's a great introduction to using simple sensors on the Raspberry Pi, and can be adapted to lots of different projects.

What you will need

As well as a Raspberry Pi with an SD card and the usual peripherals, you will also need...

Hardware

- 1 x passive infra-red sensor
- 3 x female-to-female jumper leads

Software

- Up-to-date SD card image
- Scratch 1.4

What you will learn

By creating a Santa detector with your Raspberry Pi you will learn...

- How to connect a passive infrared (PIR) sensor to the Raspberry Pi
- How to control the flow of your Scratch program by responding to the input from the sensor

1. Connect the PIR motion sensor

Before booting, connect the PIR module to the Raspberry Pi.

Using three female-to-female jumper cables, you will need to connect each of the PIR sensor's connectors to the appropriate pins on the Raspberry Pi.

Connect the top one labelled VCC on the PIR sensor to the 5V pin on the Raspberry Pi, connect the middle one labelled OUT to the GPIO 4 pin, and connect the bottom one labelled GND to a ground pin also marked GND. All shown in the following diagram...

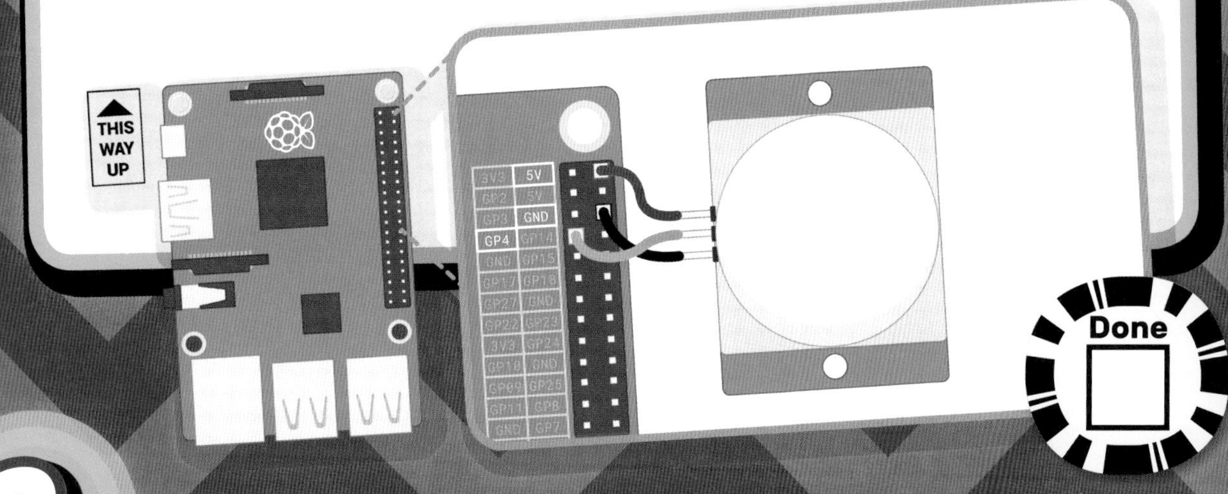

2. Test the sensor

Because we are using the GPIO pins, we need to start the GPIO server in Scratch...

- On the desktop run Scratch using... Menu > Programming > Scratch

- Once Scratch is running choose... Start GPIO Server from the Edit menu.

File Edit Share Help
Undelete
Start Single Stepping
Set Single Stepping
Compress Sounds...
Compress Images...
Show Motor Blocks
Start GPIO server

Scratch uses the Sensing blocks to check if there is any input on the GPIO pins. If there is an input, the value of the pin changes from 0 to 1. As you connected the PIR sensor to the GPIO 4 pin of the Pi, we need to monitor that.

Firstly we need to tell Scratch that GPIO 4 will be used as an input by configuring it.

- Create a broadcast message as follows...

broadcast config4in

Note

If you do not see **gpio4** on the list, make sure that the **GPIO server is running** and that you have **run the config broadcast**.

- Double-click the broadcast block to run it. You only need to do this once.
- In the Sensing block palette, click on the drop-down menu on the sensor value block and choose gpio4.
- Tick the checkbox to the left of the block to display the pin value on screen.

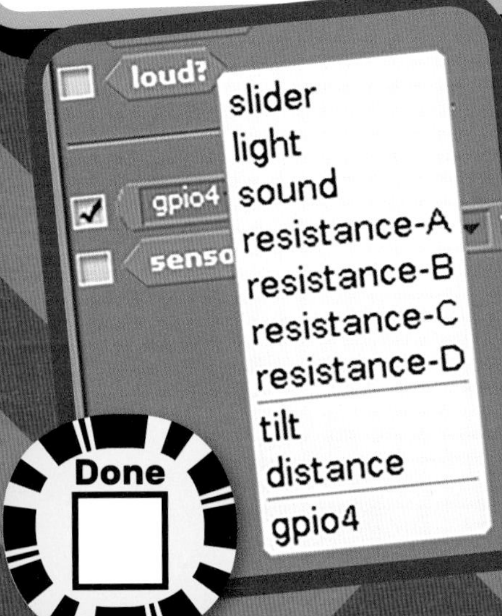

Test the PIR sensor by waving your hand in front of it. When it detects movement, the value on the screen should change from 0 to 1.

If the value doesn't change, check that the correct pins are connected.

3. Draw a sprite and add sound

Click on the Costumes tab and draw a Santa sprite. This will be displayed when the PIR senses movement.

Click on the Sounds tab and import a sound from the Electronic folder. We have used a siren called Whoop here.

4. Program what happens when the detector spots movement

Now that we have a sensor that reports when it is on or off, we can use this value to control the flow of our program.

Build the following script...

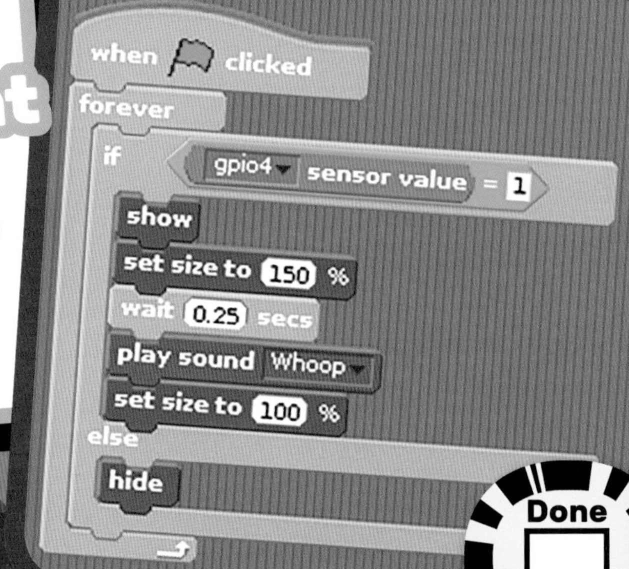

What is the the program doing?

The if statement continuously checks the GPIO 4 pin sensor value. When this changes to 1, it does the following...

- Shows the sprite
- Makes the sprite bigger
- Waits a bit
- Plays a sound
- Makes the sprite normal size

It keeps doing this as long as the sensor value is 1, i.e. when the PIR detects movement.

The else statement simply hides the sprite when the sensor value is 0.

5. Set up the detector in your bedroom!

- On Christmas Eve, set up your Pi with the sensor pointing at your bedroom door
- Connect your Pi to a huge speaker
- Make sure the sensor does not detect you in bed or you will get false positives: the alarm will go off every time you move!
- Go to sleep
- Wake up when Santa comes in and feed him mince pies and sherry!

Disclaimer!

We cannot guarantee that this alarm will not scare Santa off so that you get no presents at all, not even a wrinkled satsuma in an old sock! Sorry about that.

What next?

Other stuff you could try to take your project to the next level...

- Make the background flash
- Animate the sprite using costumes
- Change the sound ('Santa Claus Is Coming to Town' would be good!)
- Use different graphic effects instead of change size
- Display a message
- Build a support or stand for the PIR module to sit on

Spot the Difference

What's missing from the Raspberry Pi?

Timed Challenge!

Take a close look at these two Raspberry Pis... there are **12 differences** between them. You have **5 minutes** to mark all of the differences you can find on picture 2. When you have all 12, write your time in the box below – the solution is on page 78!

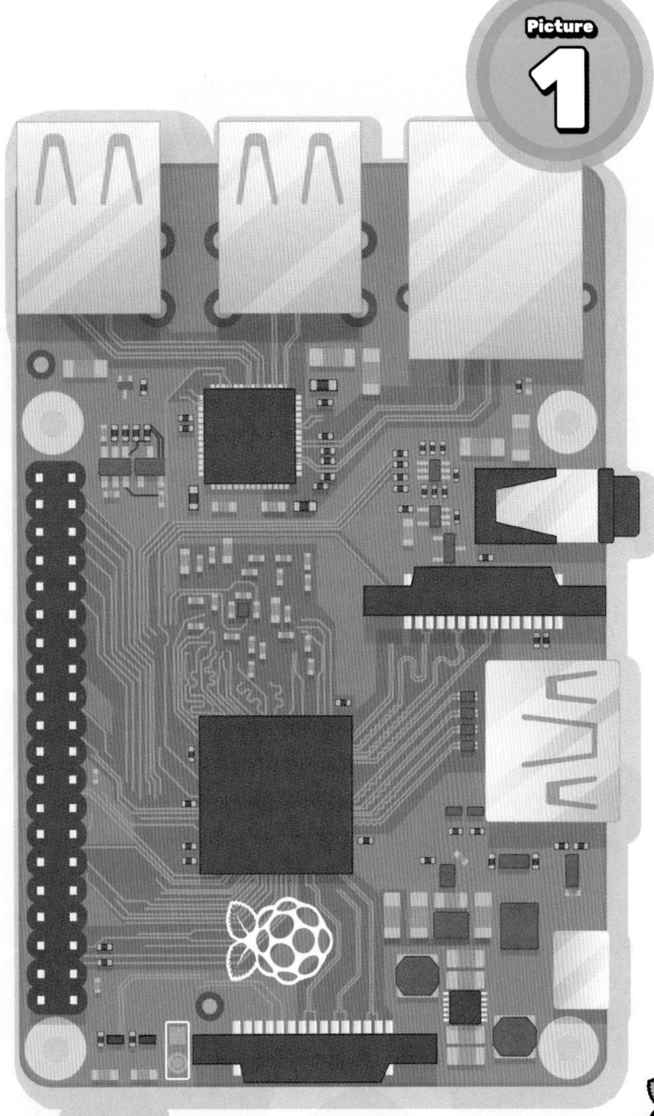

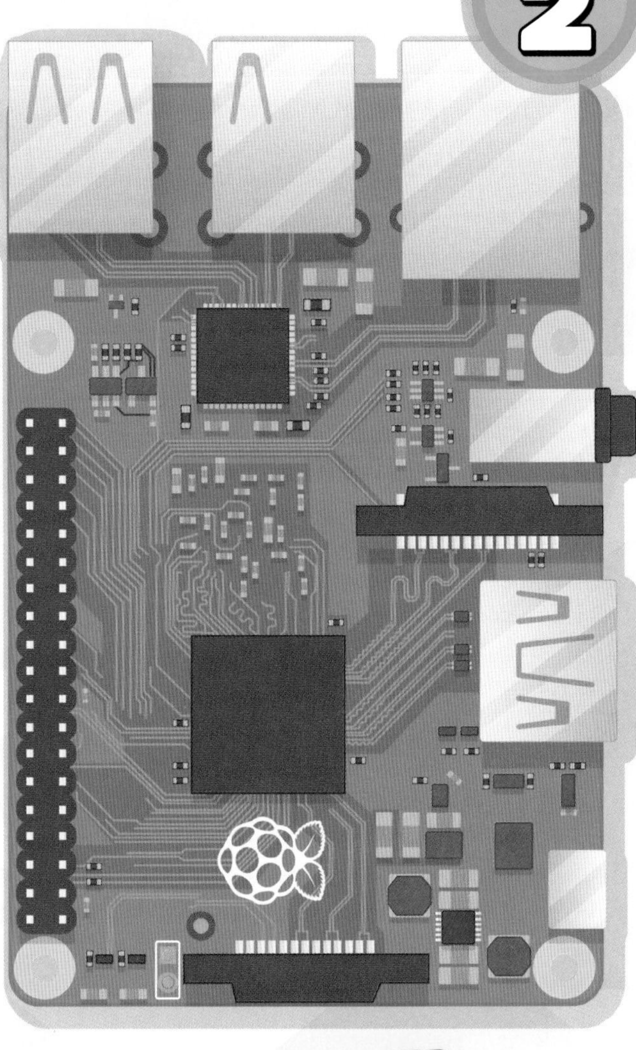

The Result!

How quickly did you manage to solve the puzzle? Write in your time below...

Big Pi Wordsearch

Find all the words in the raspberry!

Timed Challenge!

There are **17 words** in this giant Raspberry. You have **10 minutes** to find them all. Go!

- Algorithm
- AstroPi
- Computing
- Gaming
- GPIO
- Memory
- Minecraft
- Network
- Program
- Python
- Raspberry Pi
- Robot
- Scratch
- SD Card
- Sensors
- USB
- WiFi

	F	M	G	O	B				K	L	S	W	M	
S	I	W	Y	I	L	T		D	G	H	C	K	J	L
B	C	O	M	P	U	T	I	N	G	I	R	S	R	B
	S	W	A	G	C	O	O	Z	P	O	A	D	K	
		U	I	S	V	Y	W	Y	W	I	T	C		
		M	Q	F	T	Y	R	T	Q	P	C	A		
	P	G	X	O	I	R	E	O	R	H	H	R	F	
A	S	C	R	O	E	N	O	U	M	Q	G	D	F	Q
G	R	F	Z	B	F	R	C	P	L	E	H	W	D	B
T	P	J	P	G	U	D	N	L	I	T	M	Z	X	R
P	F	S	F	S	M	H	T	I	R	O	G	L	A	S
M	A	A	E	G	Y	U	H	P	E	B	N	E	R	L
R	A	V	R	F	N	F	N	F	M	O	P	O	J	W
	J	R	P	C	O	I	V	V	H	R	S	D	F	
		J	G	V	E	R	M	T	P	N	X	I		
		K	H	O	Y	N	Y	A	E	T	T	A		
				C	R	P	I	S	G	X				
						P	V	M						

Find the answers on page 78

The Result!

How quickly did you manage to solve the puzzle? Write in your time below...

Coding with Minecraft

Programming the game on the Pi!

Minecraft is a popular sandbox open-world building game. A free version of Minecraft is available for the Raspberry Pi; it also comes with a programming interface. This means you can write commands and scripts in Python code to build things in the game automatically. It's a great way to learn Python!

What you will learn

By following this project with your Raspberry Pi, you will learn...

- How to access Minecraft Pi and create a new world
- How to use the Python programming environment IDLE to connect to Minecraft Pi
- How to use the Minecraft Python API to post text to the chat window, find the player's coordinates, teleport and build structures
- How to use variables to store IDs for different types of blocks
- Experimenting with placing different types of blocks with special attributes

YOU CAN CONTROL MINECRAFT FROM YOUR PI!

What you will need

You'll need a Raspberry Pi running Raspbian, which includes Minecraft Pi by default. If you are missing the latter, however, you can download it from... **minecraft.net/en-us/edition/pi**

Tick off each step as you go!

Run Minecraft

To run Minecraft Pi, open it from the desktop menu, in Games, or type `minecraft-pi` in the Terminal.

When Minecraft Pi has loaded, click on Start Game, followed by Create new. You'll notice that the containing window is offset slightly. This means to drag the window around, you have to grab the title bar behind the Minecraft window.

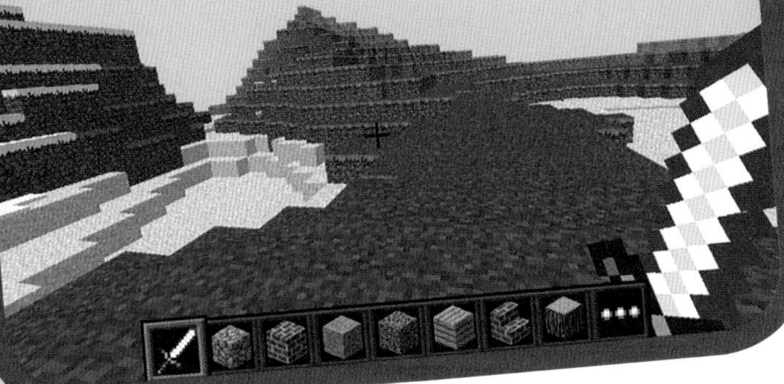

You are now in a game of Minecraft! Go walk around, hack things and build things!

USE THE MOUSE TO LOOK AROUND AND USE THE FOLLOWING KEYS ON THE KEYBOARD...

Key	Action
W	Forward
A	Left
S	Backward
D	Right
E	Inventory
SPACE	Jump
Double SPACE	Fly/Fall
ESC	Pause/ Game menu
TAB	Release mouse cursor

You can select an item from the quick draw panel with the mouse scroll wheel (or use the numbers on your keyboard), or press E and select something from the inventory.

You can also double-tap the SPACE bar to fly into the air. You'll stop flying when you release the SPACE bar, and if you double-tap it again you'll fall back to the ground.

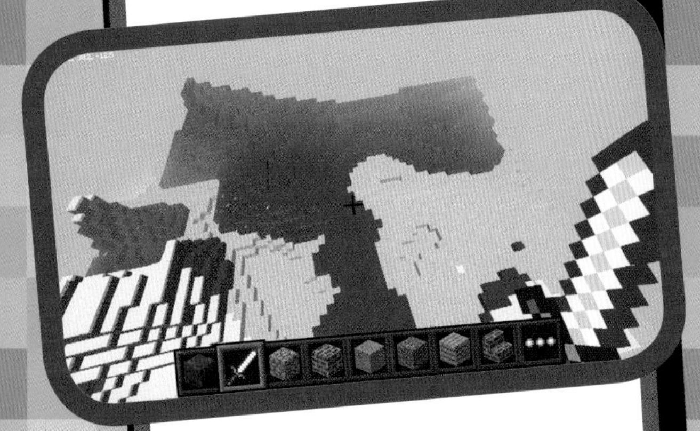

With the sword in your hand, you can click on blocks in front of you to remove them (or to dig). With a block in your hand, you can use right click to place that block in front of you, or left click to remove a block.

Done

2. Use the Python programming interface

With Minecraft running, and the world created, bring your focus away from the game by pressing the TAB key, which will free your mouse. Open Python 3 (IDLE) from the Programming menu and move the windows so they are side-by-side.

You can either type commands directly into the Python window or create a file so you can save your code and run it again another time.

If you want create a file, go to File > New window and File > Save. You'll probably want to save this in your home folder or a new project folder.

Start by importing the Minecraft library, creating a connection to the game and testing it by posting the message "Hello world" to the screen...

```
from mcpi.minecraft import Minecraft
mc = Minecraft.create()
mc.postToChat("Hello world")
```

If you're entering commands directly into the Python window, just hit ENTER after each line. If it's a file, save with CTRL+S and run with F5. When your code runs, you should see your message on screen inside the Minecraft game.

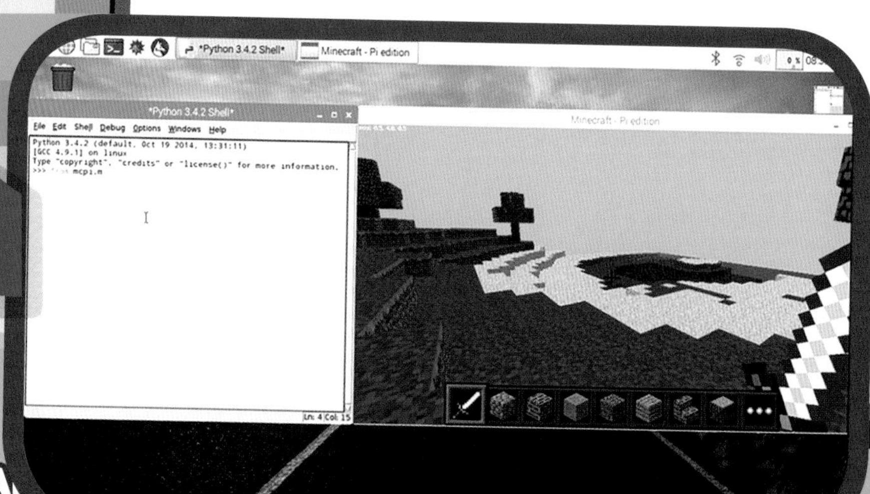

1. Find your location

Done

To find your location, type...

```
pos = mc.player.getPos()
```

`pos` now contains your location; access each part of the set of coordinates with `pos.x`, `pos.y` and `pos.z`.

Alternatively, a nice way to get the coordinates into separate variables is to use Python's unpacking technique...

```
x, y, z = mc.player.getPos()
```

Now `x`, `y`, and `z` contain each part of your position coordinates. `x` and `z` are the walking directions (forward/back and left/right) and `y` is up/down.

Note that `getPos()` returns the location of the player at the time, and if you move position you have to call the function again or use the stored location.

2. Teleport

As well as finding out your current location, you can specify a particular location to teleport to.

```
x, y, z = mc.player.getPos()
mc.player.setPos(x, y+100, z)
```

This will transport your player to 100 spaces in the air. This will mean you'll teleport to the middle of the sky and fall straight back down to where you started.

Try teleporting to somewhere else!

Done

3. Set block

You can place a single block at a given set of coordinates with `mc.setBlock()` ...

```
x, y, z = mc.player.getPos()
mc.setBlock(x+1, y, z, 1)
```

Now a Stone block should appear beside where you're standing. If it's not immediately in front of you it may be beside or behind you. Return to the Minecraft window and use the mouse to spin around on the spot until you see a grey block directly in front of you.

The arguments passed to `set.Block` are `x`, `y`, `z` and `id`. The `(x, y, z)` refers to the position in the world (we specified one block away from where the player is standing with `x + 1`) and the `id` refers to the type of block we'd like to place. `1` is Stone.

Other blocks you can try...

```
Air: 0
Grass: 2
Dirt: 3
```

Now with the block in sight, try changing it to something else...

```
mc.setBlock(x+1, y, z, 2)
```

You should see the grey Stone block change in front of your eyes!

Done

4. Block constants

You can use built-in block constants to set your blocks, if you know their names. You'll need another `import` line first, though.

```
from mcpi import block
```

Now you can write the following to place a block...

```
mc.setBlock(x+3, y, z, block.STONE.id)
```

Block IDs are pretty easy to guess, just use ALL CAPS, but here are a few examples to get you used to the way they are named...

```
WOOD_PLANKS
WATER_STATIONARY
GOLD_ORE
GOLD_BLOCK
DIAMOND_BLOCK
NETHER_REACTOR_CORE
```

5. Block as variable

If you know the ID of a block, it can be useful to set it as a variable. You can use the name or the integer ID.

```
dirt = 3
mc.setBlock(x, y, z, dirt)
```

or

```
dirt = block.DIRT.id
mc.setBlock(x, y, z, dirt)
```

6. Special blocks

There are some blocks that have extra properties, such as Wool which has an extra setting where you can specify the colour. To set this use the optional fourth parameter in `set.Block` ...

```
wool = 35
mc.setBlock(x, y, z, wool, 1)
```

Here the fourth parameter `1` sets the Wool colour to orange. Without the fourth parameter it is set to the default (`0`) which is white. Some other colours are...

```
2: Magenta
3: Light Blue
4: Yellow
```

Try some more numbers and watch the block change colour!

Other blocks that have extra properties are wood (`17`): oak, spruce, birch, etc; tall grass (`31`): shrub, grass, fern; torch (`50`): pointing east, west, north, south; and more.

Done

7. Set multiple blocks

As well as setting a single block with `set.Block`, you can fill in a volume of space in one go with `set.Blocks` ...

```
stone = 1
x, y, z = mc.player.getPos()
mc.setBlocks(x+1, y+1, z+1, x+11, y+11, z+11, stone)
```

This will fill in a 10 × 10 × 10 cube of solid Stone.

You can use the `set.Blocks` function to create bigger volumes, but it may take longer to generate!

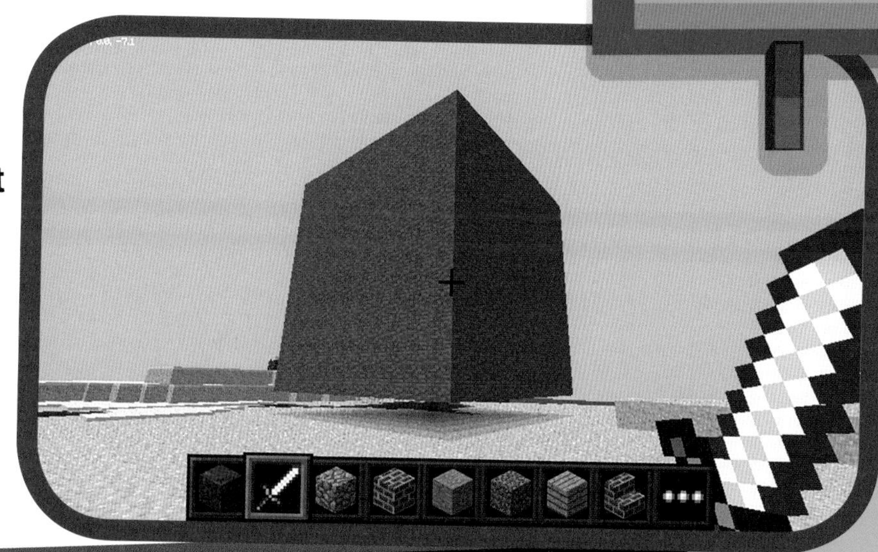

3. Playing with TNT blocks

Another interesting block is TNT! To place a normal TNT block, use...

```
tnt = 46
mc.setBlock(x, y, z, tnt)
```

However, this TNT block is fairly boring. Try applying `data` as `1`...

```
tnt = 46
mc.setBlock(x, y, z, tnt, 1)
```

Now use your Sword and left click the TNT block: it will be activated and will explode in a matter of seconds!

Try making a big cube of TNT blocks!

```
tnt = 46
mc.setBlocks(x+1, y+1, z+1, x+11, y+11, z+11, tnt, 1)
```

Now you'll see a big cube full of TNT blocks. Go and activate one of the blocks and then run away to watch the show! It'll be really slow to render the graphics as so many things are changing at once.

Done

4. Fun with flowing lava

One block that's a lot of fun to play with is flowing Lava.

```
from mcpi.minecraft import Minecraft
mc = Minecraft.create()
x, y, z = mc.player.getPos()
lava = 10
mc.setBlock(x+3, y+3, z, lava)
```

Find the block you've just placed, and you should see Lava flowing from the block to the ground.

The cool thing about Lava is that when it cools down it becomes rock. Move to another location in your world and try this...

```
from mcpi.minecraft import Minecraft
from time import sleep

mc = Minecraft.create()

x, y, z = mc.player.getPos()

lava = 10
water = 8
air = 0

mc.setBlock(x+3, y+3, z, lava)
sleep(20)
mc.setBlock(x+3,y+5, z, water)
sleep(4)
mc.setBlock(x+3, y+5, z, air)
```

YOU CAN ADJUST THE SLEEP PARAMETERS TO ALLOW MORE OR LESS LAVA TO FLOW

Done

There's plenty you can do now you know your way around the Minecraft world and how to use the Python interface.

- **Networked game**
  If multiple people connect Raspberry Pis to a local network, they can join the same Minecraft world and play together. Players can see each other in the Minecraft world.

- **API reference**
  For a more extensive documentation of functions and a full list of block IDs, visit **bit.ly/MinecraftAPI**

- **Make a game**
  Try out another resource and make a Whac-a-mole game: Minecraft Whac-a-Block.

Find links to these projects and more at... www.raspberrypi.org

Spot the Difference...

Take a trip to outer space!

Timed Challenge!

There are **12 differences** between these two pictures of the Raspberry Pi gang in space. Set a stopwatch, then mark all the differences you find on picture 2 and write in your time. And then it's time to jump to page 78 to see if you got it right!

Picture 1

Raspberry Pi in space!

Did you know there are two Raspberry Pis on the International Space Station? Learn more at **astro-pi.org**!

The Result!

How quickly did you manage to solve the puzzle? Write in your time below...

Find the answers on page 78

Babbage versus Bugs

Code our exciting Space Invaders clone!

Download the project

Go to **goo.gl/PyzcFA**, press the green 'Clone or download' link on the right and select 'Download ZIP'.

Set up the files

Double-click on the downloaded zip file in your Downloads folder and click the 'Extract files' icon, then 'Extract'. Now click through to the **bugs** folder to find the project.

Look at the code

Right-click on **bugs.py** and open with Thonny or a text editor. You'll find the following code listing. Check out the code, then go to the end of the project (page 45) see how to play it!

It's much more rewarding to write the code yourself! Press the menu button, go to Programming and select Thonny. Open a new file and write out the code as you see it here. When you're done save it in the 'bugs' project folder you downloaded with '.py' at the end of the file name.

bugs.py

Set sizes

Let's set the resolution and name our game.

```
from random import randint as rand

WIDTH = W = 640
HEIGHT = H = 480
TITLE = "Babbage vs Bugs"

P = [(0,1,240), (-1,0,112), (0,1,32), (1,0,224), (0,1,32), (-1,0,224)]
F = [lambda x, y : 0,
     lambda x, y : y<1,
     lambda x, y : y==1,
     lambda x, y : y>1,
     lambda x, y : x&1,
     lambda x, y : (x^y)&1,
     lambda x, y : 1]
```

```
class Bug(Actor):
	def __init__(s, pos, kind):
		super().__init__("blank", pos)

		s.kind = kind
		s.life = kind*2
		s.time = 0

	def update(s):
		s.time -= 1
		s.x += P[state.pc0][0]
		s.y += P[state.pc0][1]

		c = s.collidelist(state.beams[0])
		if c >= 0:
			state.beams[0][c].h = 1

			s.life -= 1
			if s.life == 0:
				state.score += s.kind*10

			s.time = 5

		if rand(0, 399) == 0:
			state.beams[1].append(Beam(s.pos, 3))

		s.image="bug"+("s" if s.time>0 else str(s.kind))+str(state.pc1>>3&3)

class Star(Actor):
	def __init__(s):
		super().__init__("star", (rand(0, W-1), rand(0, H-1)))

		s.v = rand(1, 3)

	def update(s):
		s.y -= s.v

		if s.y < 0:
			s.x = rand(0, W-1)
			s.y += H

class Beam(Actor):
	def __init__(s, pos, v):
		super().__init__("beam", pos)

		s.v = v
		s.h = 0

	def update(s):
		s.y += s.v

class Player(Actor):
	def __init__(s):
```

YES! WE'VE HIT A PESKY BUG! WE NEED TO REMOVE A LIFE WITH s.life -= 1

```
        super().__init__("blank", (W/2, H-64))

        s.time0 = 0
        s.time1 = 0
        s.life = 5

    def update(s):
        s.time0 -= 1
        s.time1 -= 1

        dx = (3 if keyboard.right else 0)-(3 if keyboard.left else 0)

        s.x = max(32, min(W-32, s.x+dx))

        c = s.collidelist(state.beams[1])
        if c >= 0 and s.time0 < 0:
            state.beams[1][c].h = 1

            s.life -= 1
            s.time0 = 5

        if keyboard.space and s.time1 < 0:
            state.beams[0].append(Beam(s.pos, -5))
            s.time1 = 15

        s.image = "bab"+("s" if s.life > 0 and s.time0 > 0 else str(state.
pc1>>4&1))

class State:
    def __init__(s):
        s.bugs = []
        s.beams = ([], [])
        s.stars = [Star() for s in range(30)]

        s.player = Player()

        s.score = 0
        s.space = 0

        s.wave = 0

    def update(s):
        if len(s.bugs) == 0:
            for y in range(3):
                for x in range(7):
                    s.bugs.append(Bug((W/2+x*60-180, y*60-180), 2 if
F[min(s.wave, 6)](x, y) else 1))

            s.pc0 = 0
            s.pc1 = 0
            s.wave += 1

        for a in s.all():
            a.update()
```

Update me

The **update** method is where we check for button presses to move Babbage left and right, and collisions to see if he's been hit by a bug.

Classy

The class **State** sets up our game, initialising things like bugs, our player, and the scoreboard.

```
        s.bugs = [b for b in s.bugs if b.life > 0]

        s.beams = ([b for b in s.beams[0] if b.y > -64 and not b.h],
                   [b for b in s.beams[1] if b.y < H+64 and not b.h])

        s.pc1 += 1
        if s.pc1 == P[s.pc0][2]:
            s.pc0 = 2 if s.pc0==5 else s.pc0+1
            s.pc1 = 0

    def all(s):
        return s.stars+s.beams[0]+s.beams[1]+s.bugs+[s.player]

    def over(s):
        return s.player.life <= 0 or len(s.bugs) and max([b.y for b in s.bugs])
> s.player.y - 50

state = State()

def update():
    global state

    if state.over():
        if keyboard.space and not state.space:
            state = State()
    else:
        state.update()

    state.space = keyboard.space

def draw():
    screen.clear()

    for a in state.all():
        a.draw()

    for i in range(state.player.life):
        screen.blit("life", (6+i*32, H-26))

    screen.draw.text(str(state.score), bottomright=(W-8, H-3), fontname="consola",
fontsize=20)

    if state.over():
        screen.blit("dark", (0, 0))
        screen.draw.text("GAME OVER", center=(W/2, H/2), fontname="consola",
fontsize=100)
```

GAME OVER

Game credits:
Eben Upton, Laurence van Someren, Sam Alder

4. Run the game!

Press the menu button, select Accessories, and then Terminal. Type: cd Downloads/Annual-2018-master/bugs. Now type pgzrun bugs.py (or the name you gave your own file). Enjoy!

Space Maze

Take a trip to outer space!

Timed Challenge!

The spaceman's teddy bear has got stuck in the centre of this space maze. Can you find the **correct pathway** to rescue him, then escape to the space station? We're giving you just **2 minutes** to do it... ready, teddy, GO!

Find the answers on page 78

The Result!

How quickly did you manage to solve the puzzle? Write in your time below...

____

DID YOU KNOW?

THERE ARE TWO PIS ABOARD THE INTERNATIONAL SPACE STATION! LEARN MORE AT...

WWW.ASTRI-PI.ORG

Raspberry Pi Personality Test

Choose an answer to each question to find out what makes you tick!

1. What are your Christmas plans for your Raspberry Pi?

A. Build a Santa detector
B. Teach it to fire darts at your sister
C. Leave it on the floor in the dark so Dad steps on it in bare feet

2. Which of these is the best Raspberry Pi project?

A. Hamster disco
B. Parent detector
C. WhooPi Cushion

3. What would make your Raspberry Pi even better?

A. An elegant moustache
B. Terrifying bat wings
C. A laser turret

4. Your Raspberry Pi broke. How did it happen?

A. Plummeted out of a tree
B. Dropped into a school dinner
C. Fell in the toilet

5. You have a terrible Raspberry Pi nightmare. What happened?

A. Killer robots
B. Automatic homework-giving machine
C. Healthy eating program, with lots of vegetables

6.

What do you think is the best name for a Raspberry Pi robot?

A. Alfonse
B. Mr Fluffypants
C. Murderborg

7.

You need to hide your Raspberry Pi from your brother. Where do you conceal it?

A. In the fruit bowl
B. In your stinky gym kit
C. In Mum's underwear drawer

8.

What robot would you most like to build with your Pi?

A. Robo-shark. With lasers.
B. Robo-dog
C. Robo-butler

Scores

Each answer you give earns you points! **Answer A = 1 point, B = 2 points and C = 3 points**. Flip the book over and find out how you scored at the bottom of the page!

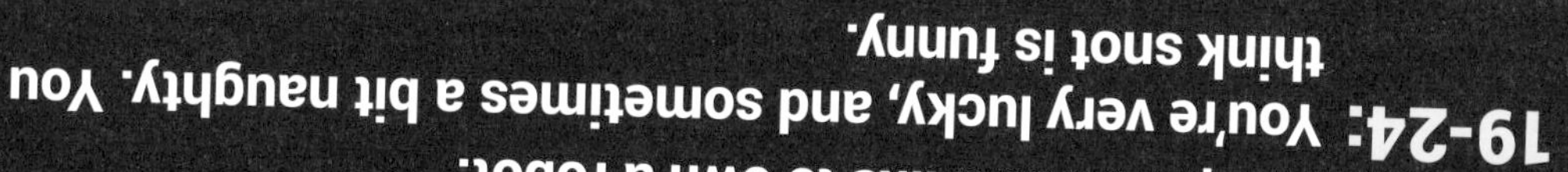

Here are the scores on the doors – find out what kind of Raspberry Pi personality you have!

8-13: You are a glittering Christmas star. Your favourite part of Christmas is opening presents with your family.

14-18: If you could be any monster, you'd probably be a vampire. You'd like to own a robot.

19-24: You're very lucky, and sometimes a bit naughty. You think snot is funny.

Wheel of Pi

Round and round – find the hidden phrase!

Timed Challenge!

This is the Wheel of Pi, and it hides a special phrase! Skip every other letter to decipher the code – you have just **5 minutes** to crack it! The pictures might help you a bit!

N H E P C I R C A O F D T E M W I I T

Find the answers on page 78

The Result!

How quickly did you manage to solve the puzzle? Write in your time below...

____

THE ALMOST CERTAINLY TRUE STORY OF

RASPBERRY PI

THE ADVENTURE THAT STARTED A DIGITAL REVOLUTION!

MANY YEARS AGO...

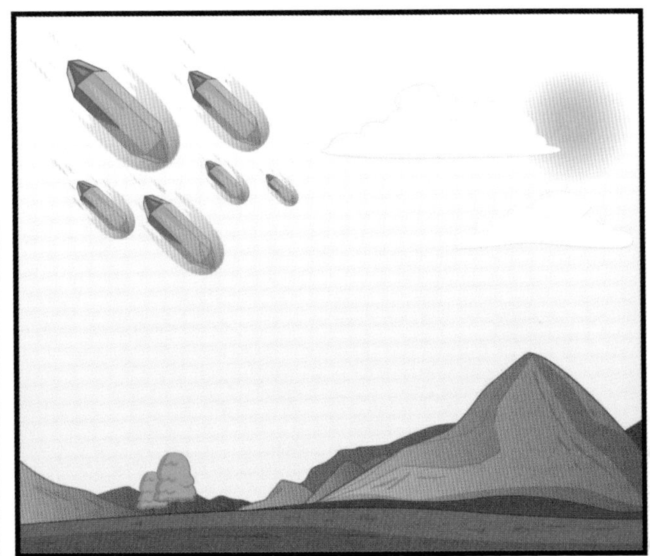

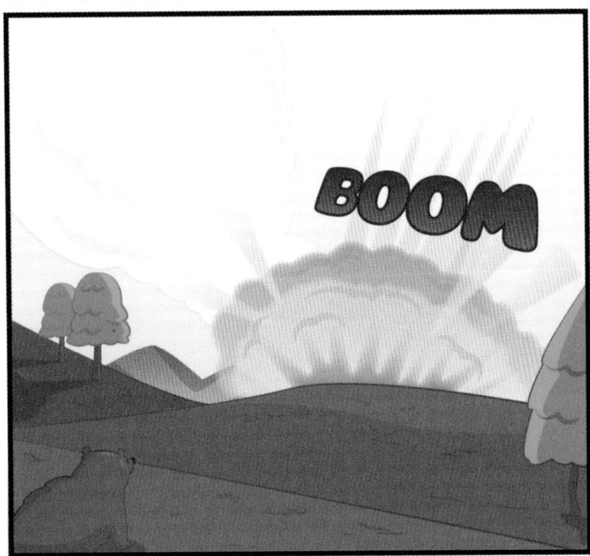
BOOM

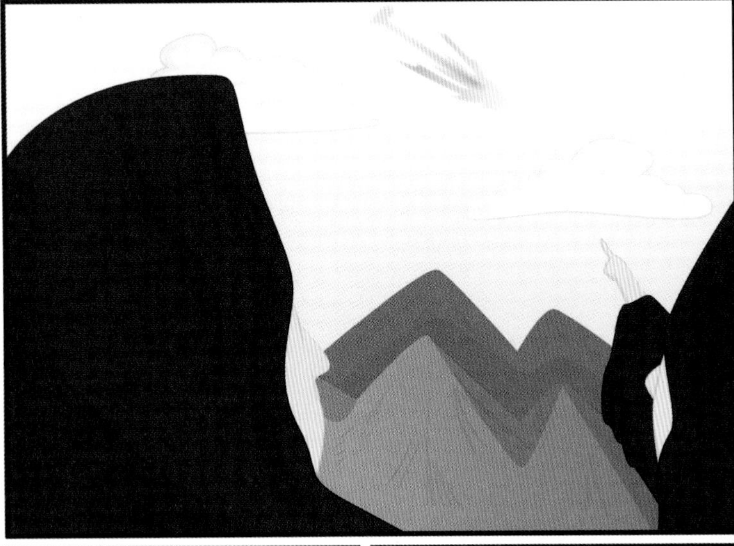

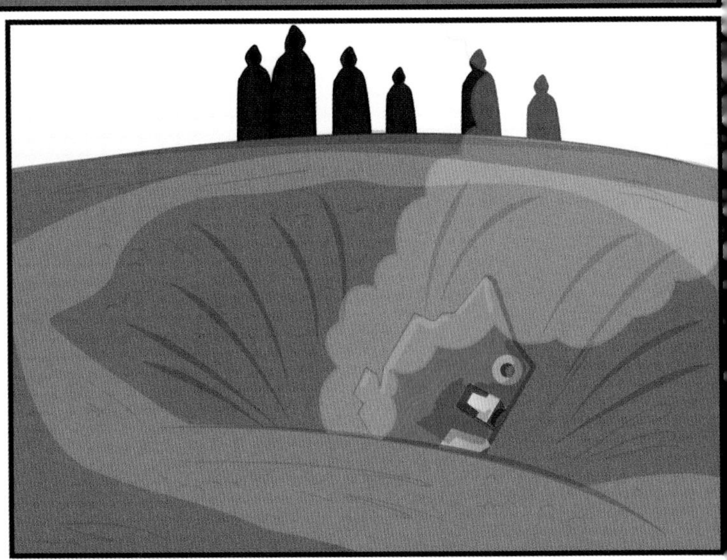

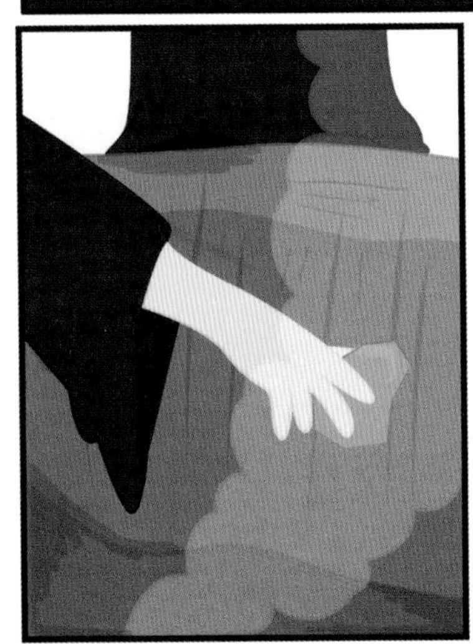

Behold—
the final
shard.

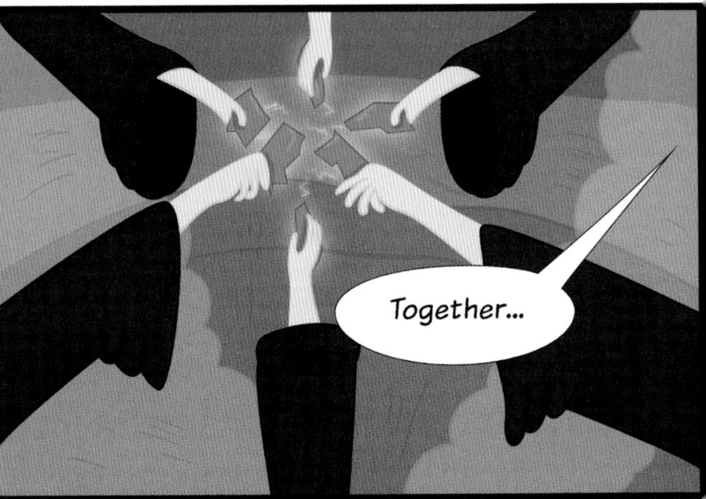
Together...

AT LAST!

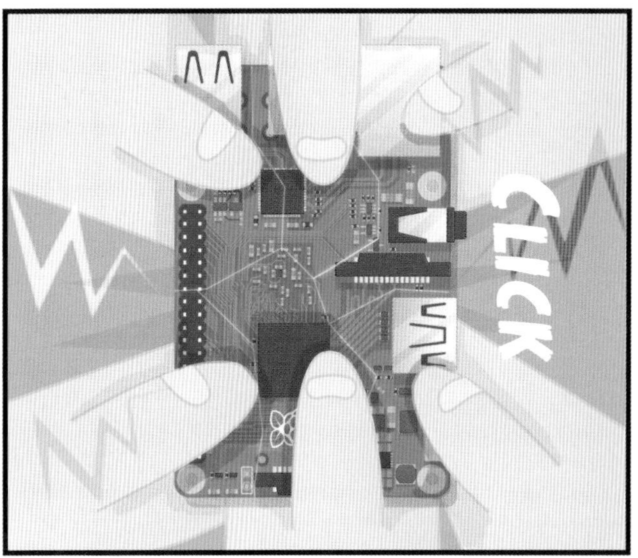
CLICK

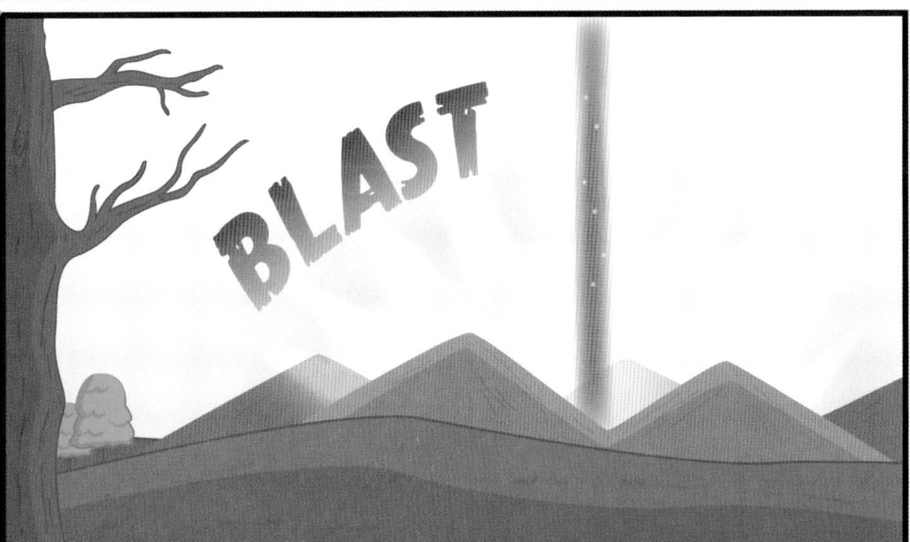
BLAST

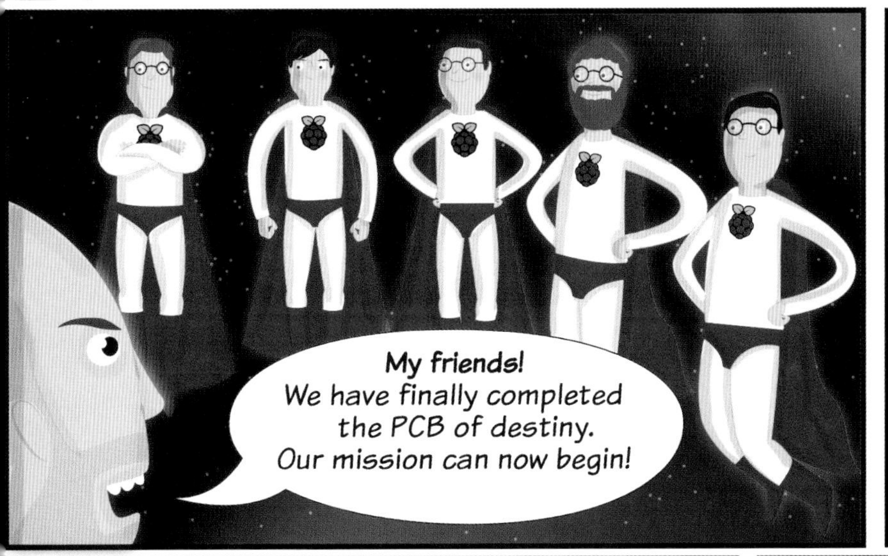
My friends!
We have finally completed the PCB of destiny.
Our mission can now begin!

With this sacred Pi, we can create a world of knowledge, peace, joy, and delicious code!

MUHAHAHAHA!

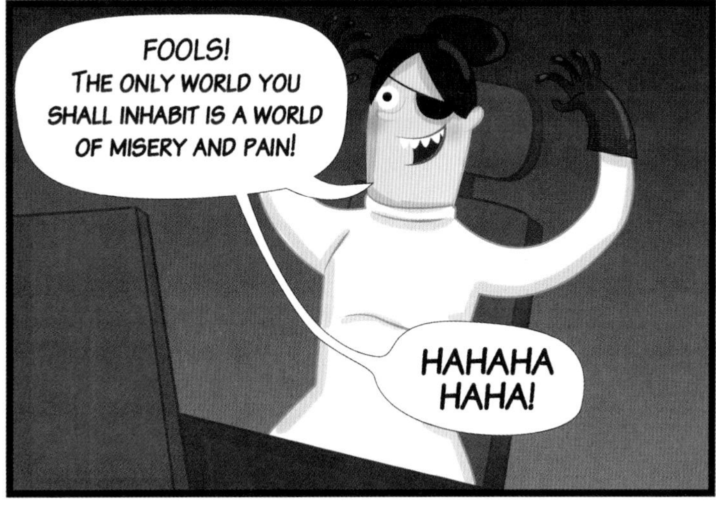
FOOLS!
THE ONLY WORLD YOU SHALL INHABIT IS A WORLD OF MISERY AND PAIN!
HAHAHA HAHA!

FIVE YEARS LATER...

Our vision has been realised. We really did it!
Hell Yeah!

Every child in the world has been empowered by the wonderous might of Raspberry Pi, every-

Sir, sir!

The Professor sent me. There's some kind of life-threatening emergency!

TO THE LAB.

DOWN IN THE PI LAB

Hello, sir.
What's the word, Professor?

UFO approaching atmosphere, Sir.
I don't like this one bit!

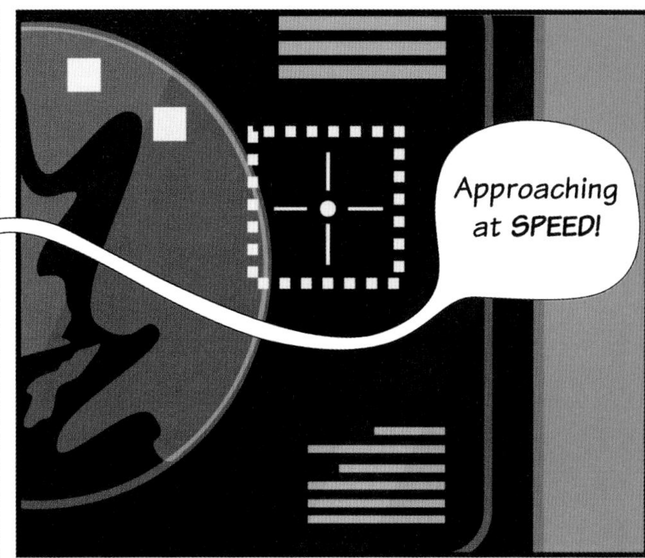
Approaching at SPEED!

BEEP! BEEP!
INCOMING COMMUNICATION!

Patch us in.

Greetings, Raspberry goons! I'll keep this brief so your puny minds can keep up!

My name is **Doctor Bluescreen.** I am here to upload your brainy children into a giant spreadsheet and use their knowledge for a variety of evil deeds.
Starting now!

ZAP

UH-OH.

Holy GPIO! Doctor Bluescreen really is uploading the children!

I must assemble **the Founders** once again to defeat this great evil.

FOUNDERS ASSEMBLE !!!

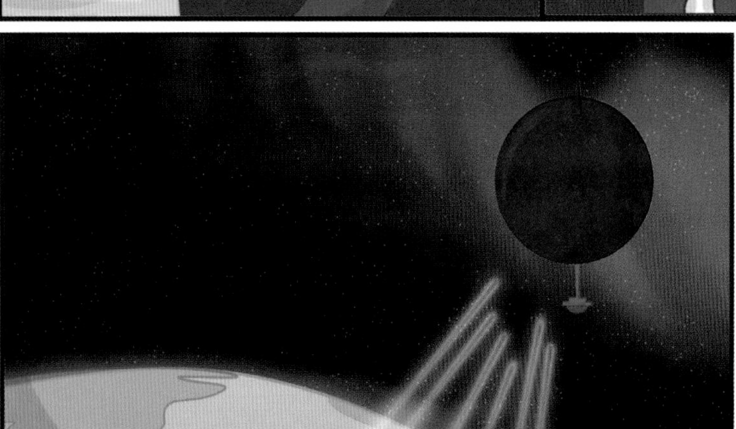

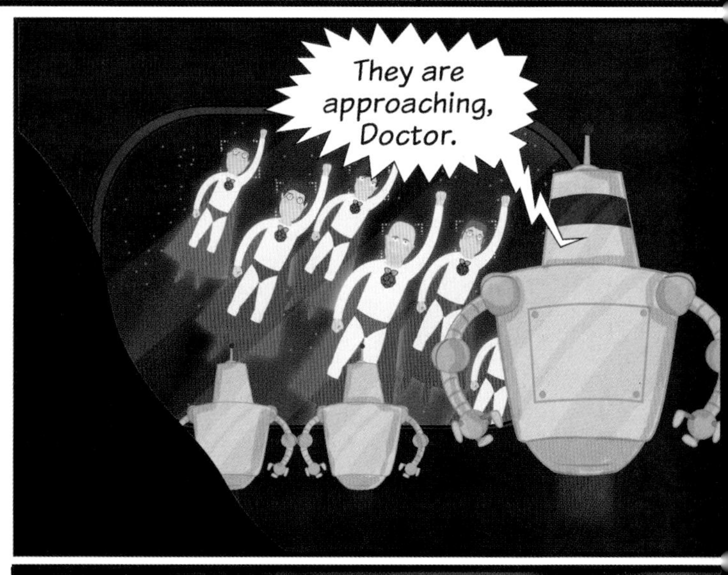
They are approaching, Doctor.

Excellent. The fools have flown straight into my sticky-icky web. Hahahah!

There she is! Time to put an end to this madness.

IT'S TIME TO
CODE.

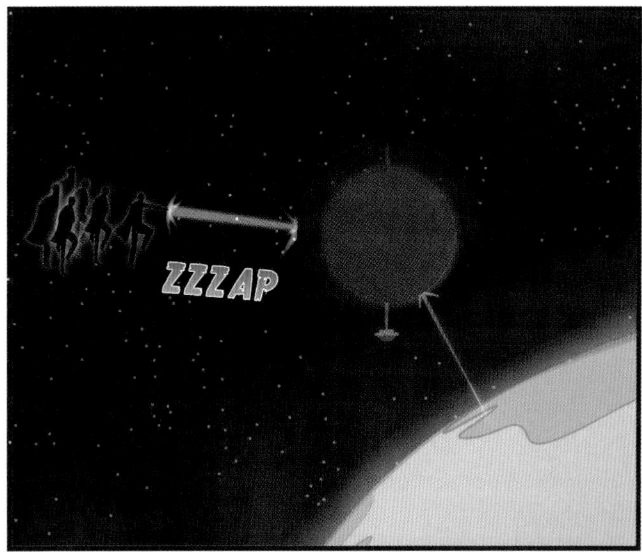
ZZZAP

Hmmm, her shields are stronger than we thought.

Haha! Your pathetic code beams are no match for my mighty orb!

We must merge our code into one beam of pure energy!

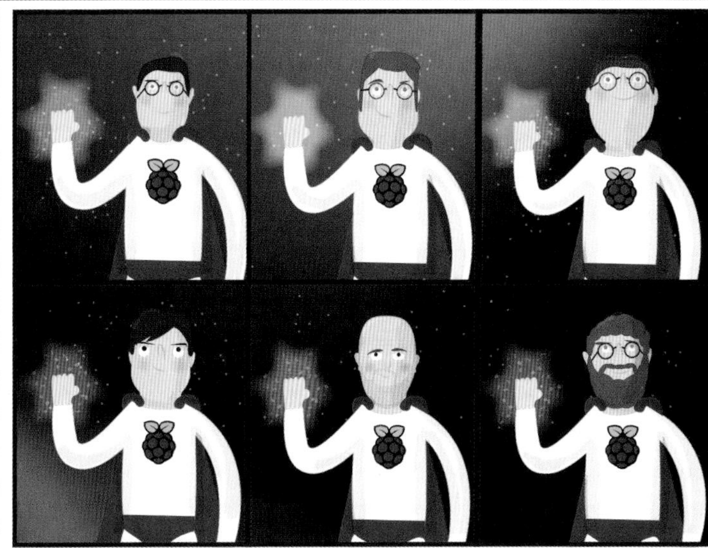

BLAST!

Argh! Still nothing! Her shields are resistant to our code!

Enough of this!
It's time to fry these
fools. There is much
evil to be done.

Release the
VIRUS

CREAK!

HUH?

OH NO!

What's...
happening? I
can't...move!

IT'S A TRAP!

BACK AT THE PI LAB
Come in, come in, do you read me?

It seems they've been trapped by some kind of virus!

WHAT THE HELL ARE WE GOING TO DO?!

Pull yourself together, man!
SLAP!

Now think, there must be something we can do.

Thanks, I needed that.

Well, legend has it that there is an ancient computer that can be used in times of dire need. But that's just some legend from 1985!

The computer could be anywhere in the world! how could we ever find it in time to save them?!

Do you mean this old thing?

THAT'S IT!

Come on, let's hack this thing back to life and save the WORLD!

Doctor, all the children are uploaded into the spreadsheet and saved to your desktop.

Excellent! Let's destroy those Raspberry fools and be on our evil way.

friends... I grow week

If this is our end, let it be known that... we fought gallantly, even if...our efforts were futile.

AT THE PI LAB
EUREKA!
It's alive!

CLICK
COOOOOOL

BUBBLE
BUBBLE

SPLASH

WHOOOOSH

It's...
MECHA
-BABBAGE

ZZZZAP!

Our shields have been disabled! The children are downloading back to earth!

IMPOSSIBLE!

We're free!
Let's CODE!

ZZZZAP!

WE'RE GONNA BLOW!

N

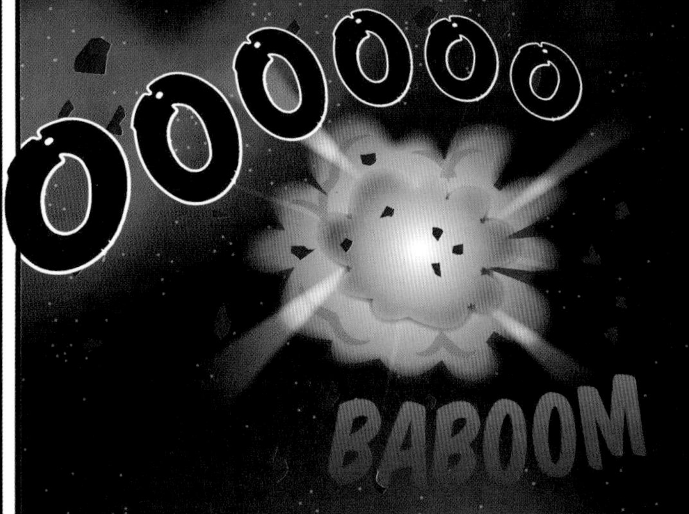
OOOOOO
BABOOM

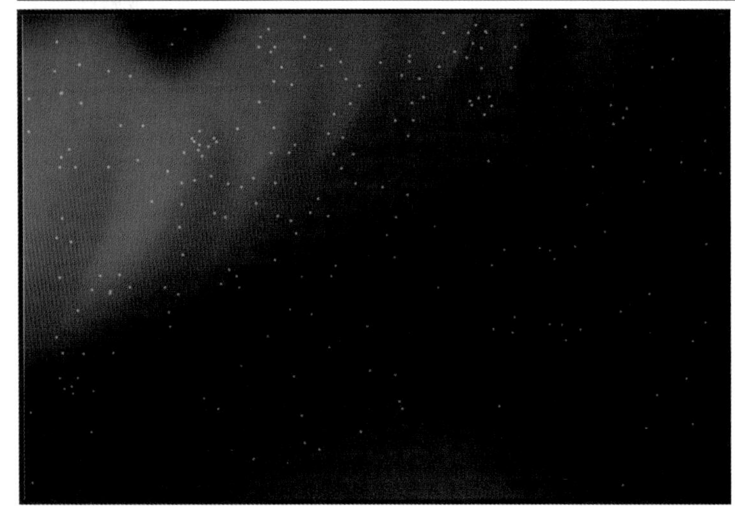

WOOHOO!

Thank you, Mecha-Babbage! How can we ever repay you?

You have freed me from my watery tomb—I can now return to my Mecha-family. That is thanks enough, my tiny friends.

See ya, Babbage.
Such a nice guy.

BACK ON EARTH
Well—we did it. The earth is safe once again!

And what have we learned from our little adventure?

That all life is a precious and delicate gift that must be nurtured and loved like a flower?

Err...yeah, that too! But the main lesson here is that...

RASPBERRY PI RULES!

Python Project

Turtley Amazing

Move from Scratch to Python with the Turtle!

In this project you will take your first steps with the programming language Python to draw shapes, patterns, and spirals. You will use a module named Turtle. Along the way you will learn how to think in sequences, and use loops to repeat a sequence. This is a great stepping stone from a visual programming language like Scratch to the text-based environment of Python.

What you will learn

By making patterns with Python code, you will learn...

- To take your first steps with the Python programming language
- How to draw lines with Python Turtle
- How to make turns
- How to change the pen colour
- How to use loops to repeat some instructions and create shapes
- How to use more loops to create impressive spiral patterns

What you will need

With your Raspberry Pi set up and an installation of Raspbian, you have everything you need for this project.

LEARN HOW TO CONTROL ME USING EASY PYTHON CODE!

Note

Make sure you don't call your Turtle projects **turtle.py** - that's the name of the Python library you're importing!

Tick off each step as you go!

1. Is it art, maths, or computer science?

Have a look at the image to the right. How would you describe it? Is is art, maths, or computer science?

It's a computer-generated image, but making it requires an understanding of art, maths, and computer science. Let's see how you too can make images just like this.

2. Drawing a line

The image above is made up of lines and only lines! To get started, you need to know how to draw a line using a little bit of Python code. Below is some code we would like you to write into Python – see the note box for options for experimenting with Python. Type in the code and then Run it to see what happens.

```
from turtle import Turtle, Screen

turtle = Turtle()
screen = Screen()

turtle.forward(100)
```

1. Click on Run to see the code working.
2. Now try changing the number in the line `turtle.forward(100)`; click on Run again and see what happens.

Note

You can run Python code in a browser using online services like Trinket (**trinket.io**), or load up Python on your Raspberry Pi. You will find the latest versions in the **Programming** menu.

3. Turning

You've used code to draw a line. Good work! Now let's try making the turtle turn around. To do this you need to instruct the turtle not only to move forward, but also to turn right or left.

```
from turtle import Turtle, Screen

turtle = Turtle()
screen = Screen()

turtle.forward(100)
turtle.right(90)
turtle.forward(100)
```

1. What do you think will happen in the code above? Click on Run to see if you were right.

`turtle.right(90)` turns the cursor 90 degrees right. You can also turn left with `turtle.left(90)`. To change the amount that the cursor turns, simply change the value of degrees.

2. Complete the square shape you've started by adding more lines of code and press Run. Keep trying until you get it right.

Done

Challenge

Try to complete each of the challenges below...

- Draw a rectangle: two of the four sides need to be longer
- Draw a triangle: how many degrees do you need to turn?
- Draw a cross: backward and forward work well together
- Draw a circle: what happens if you turn lots?

4. Changing colours

The default colour for the pen used by the turtle cursor is black, and the default background colour is white. You can change the colours to make your shapes look even better.

1. Look at the code below. It contains three variables called `R`, `G`, and `B`.

```
from turtle import Turtle, Screen

turtle = Turtle()
screen = Screen()
screen.colormode(255)
R = 255
G = 255
B = 0

screen.bgcolor((R, G, B))
```

Variables are a way of storing a value and giving it a name. For instance, there is a variable name `R` with a value of `255`.

2. Run the code and see what happens.

3. Try changing the values of the three variables, and see what happens. (Note: the maximum value is 255, and after this there will be no effect.) What do you think R, G, and B represent? You can change the value of your variables either by setting them to a new value, or by increasing and decreasing them.

4. You can change the colour of the turtle as well. Run the code below to see what happens...

```
from turtle import
Turtle, Screen

turtle = Turtle()
screen = Screen()
screen.colormode(255)
R = 255
G = 0
B = 124

turtle.color((R, G, B))
turtle.forward(100)
turtle.right(120)
turtle.forward(100)
```

Challenge

Try to complete each of the challenges below...

- **Complete the triangle above with a colour of your choice**
- **Draw a square with sides which are four different shades of red**
- **Draw a cross made of four different colours**

5. Repetition

Repeating lines of code is one of the fastest ways to get something done. Quite often in computer science, it makes more sense to repeat lines of code rather than write out another set of instructions. For example, the square you created earlier uses the same two instructions four times. Rather than writing them out four times, you could write them out once but add an instruction to repeat them.

In Python there are two types of loops that you are likely to use: a while loop and a for loop. If you want a section of code to repeat forever, or until a condition is set, then a while loop might be best. If you want to loop for a set number of times, then a for loop is preferable.

1. Here, we have used a while True loop. This means that the code inside the loop (i.e. the code which is indented) will repeat forever. You can try to see what it does, but remember it will loop around forever!

```
from turtle import Turtle, Screen

turtle = Turtle()

while True:
    turtle.forward(1)
    turtle.right(1)
```

This type of loop is not going to be very useful for drawing shapes with Turtle where you want to be more precise.

2. In this example, a `for` loop has been used. Press Run to see what happens.

```
from turtle
import Turtle,
Screen

turtle = Turtle()
screen = Screen()

turtle.penup()

for i in range(8):
    turtle.write(i)
    turtle.forward(20)
```

A `for` loop repeats instructions a set number of times, in this case 8 times. A `for` loop has an associated variable (called `i` here). In this example, `i` starts from `0` and increases by `1` each time. Let's apply this to the code to draw a square...

```
from turtle import
Turtle, Screen

turtle = Turtle()

for i in range(4):
    turtle.forward(100)
    turtle.right(90)
```

Done

3. In the code in step 2, the turtle has been asked to repeat two instructions four times to make a square.

4. Once you have created one shape using a loop, you can repeat the shape again and again by putting it inside another loop. This is a great way to draw spirals. Adapt your code by making it look like this...

```
from turtle import Turtle, Screen

turtle = Turtle()

for i in range(30):
  for i in range(4):
      turtle.forward(100)
      turtle.right(90)
  turtle.right(25)
```

A spiral can be made by turning a small degree and then moving forward a small amount. The section of code for making a square is inside another `for` loop that repeats it 30 times, each time turning the cursor 25 degrees to make a pleasing spiral shape.

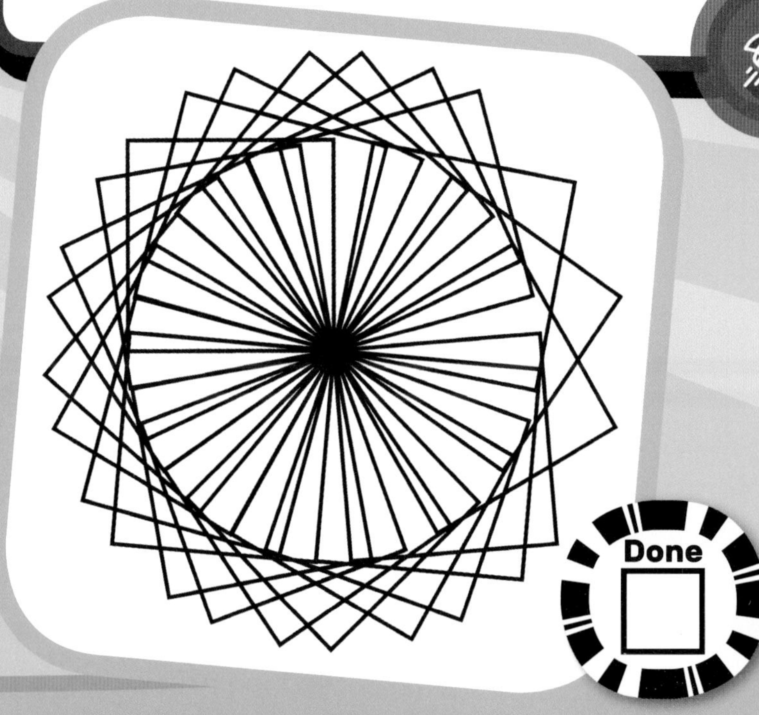

Done

Challenge

Try to complete each of the challenges below...

- Can you alter the `for` loop so that it draws a more interesting spiral using one of the shapes you made earlier, like a triangle or circle?
- Adding a few extra lines where you alter the variables `R`, `G`, and `B` would allow you to make a multicoloured spiral. Have a go at creating a rainbow spiral.
- Draw a circle: what happens if you turn lots?

Other things to try...

Take your Raspberry Pi tricks to the next level with these ideas!

- Learn how to use functions to draw snowflakes using Turtle
- Create interactive stories using lists in Python with the Storytime resource found on the Raspberry Pi website
- Take your first steps controlling physical objects with Python and a Raspberry Pi
- Become a Python turtle expert by continuing with the advanced tutorial of Turtley Amazing!

Visit...
www.raspberrypi.org

Make a WhooPi Cushion

Bring the fart gag up to date with Pi!

In the bad old days before TV and computers, the most popular family entertainment was the whoopee cushion, a tooty balloon made from an unwashed pig's bladder. This was inflated and hidden under grandad's chair cushion – when he sat down it 'PARPED!' loudly, making him jump into the air and his false teeth fly out. It was the best thing ever (especially when the dog caught the teeth and ran about wearing them and grinning). This project brings the whoopee cushion up to date: no bladder; no need to blow it up; and you can add whatever noises you want!

What you will need

To build your very own farting machine you are going to need...

- 2 paper plates
- A washing-up sponge
- Kitchen foil
- Sticky tape
- Glue or double-sided tape
- 2 female header wires (that fit on the GPIO pins)
- 2 lengths of thin, insulated wire
- A speaker (the louder the better!)

HEE-HEE! EVERYONE WILL THINK GRANDAD HAS FARTED!

What you will learn

Apart from learning which members of your family have a sense of humour, you will also learn...

- How to create Raspberry Pi sensors with household objects
- How to code simple programs in Python to carry out tasks
- How to use the Terminal

Tick off each step as you go!

1. Making the WhooPi Cushion

1. Tape or stick squares of foil on the middle of the "eating" side of each plate. These are your contacts – when they touch, they'll make a circuit.
2. Strip the end of one of the long wires and tape it to one of the squares of foil. Make sure it makes good contact with the foil. Do the same with the other wire and the other plate.
3. Chop the sponge up into cube chunks and glue them around the foil on one of the plates – this will stop the foil squares touching each other until someone sits on the plate. It should look something like this...

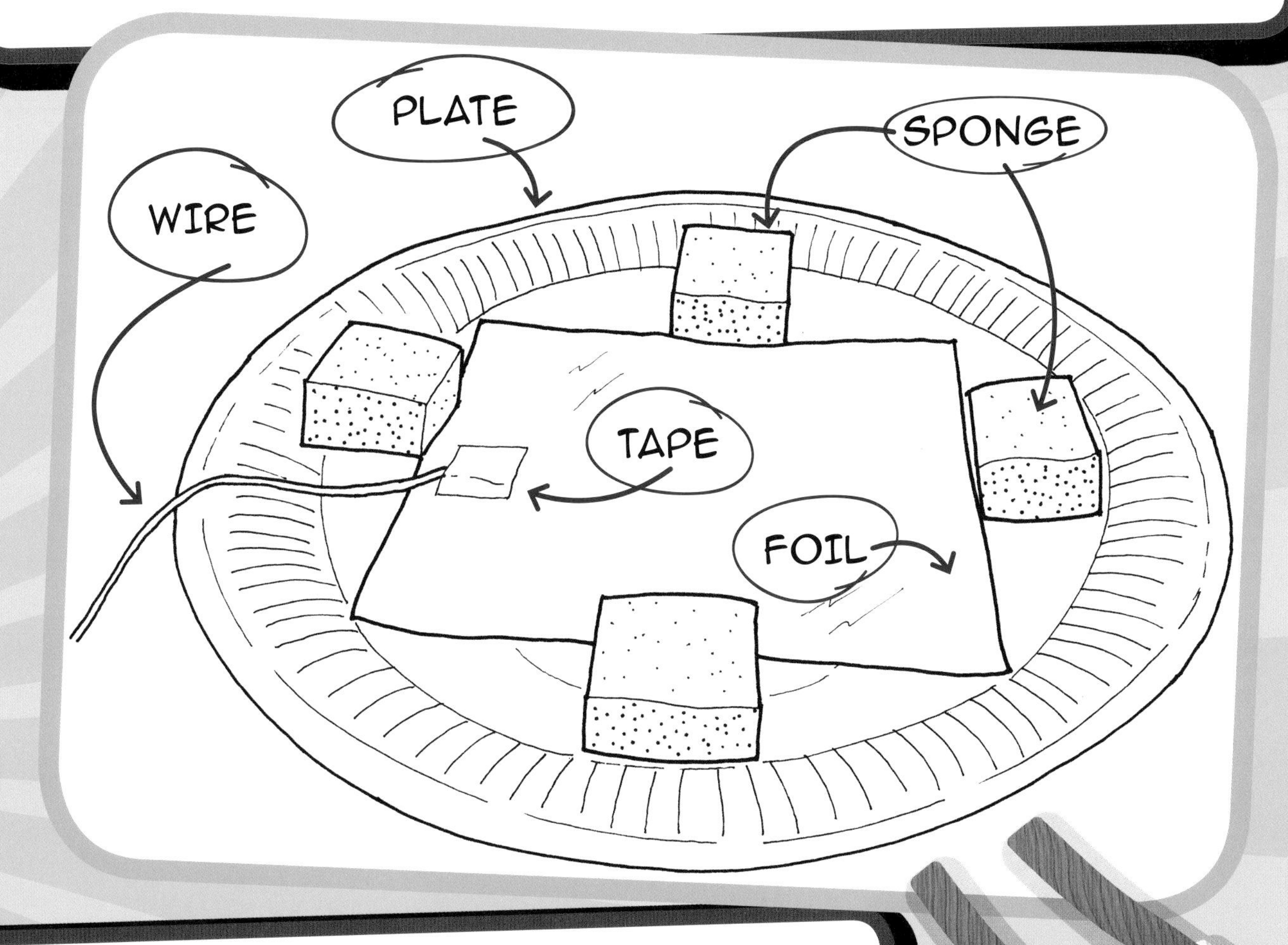

4. Put the two plates together so that foil squares are on the inside and facing each other. Tape them together.
5. You now have a 'cushion' made of two plates with two connecting wires coming out. Later we will connect these wires to the GPIO pins on your Raspberry Pi: one to a ground pin and the other to a pin that we will use to detect when the circuit is made. For this we will use the header wires.

6. Strip the ends of the connecting wires and attach each one to a female header lead. One way is to cut off one end of the header lead, strip it, twist it to the long connecting wire, and then insulate the join with tape – but do whatever works best for you.

CUT HERE

STRIP END AND JOIN IT TO LONG CONNECTING WIRE

It's now time to hook up the Pi to your finished WhooPi cushion!

2. Connect the WhooPi Cushion to the Pi

Note that if you have an older Raspberry Pi model you'll only have 26 pins, but they have the same layout.

1. Plug one header lead (it doesn't matter which) onto a ground (GND) pin on the Pi...

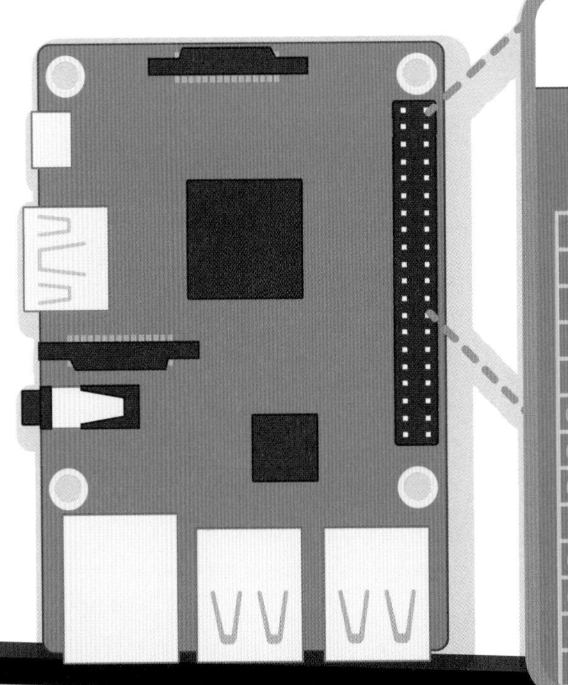

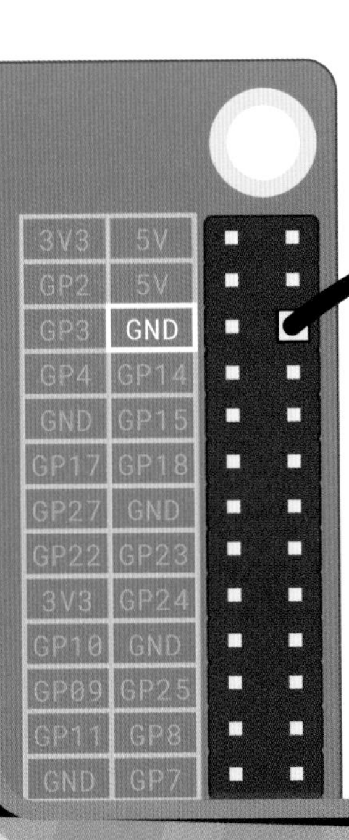

2. Plug the other wire into GPIO pin 2...

You can damage your Raspberry Pi if you do not use the GPIO pins correctly. Stay away from the 5V pins!

WARNING!

That's the hardware complete, now for the software! We are going to use Python – don't worry if you've not used it before, just follow the instructions and you will pick it up.

You will be using the command line to type stuff in. To do this you will need to open a Terminal window by clicking on the screen-like Terminal icon, three along from the menu icon on your desktop...

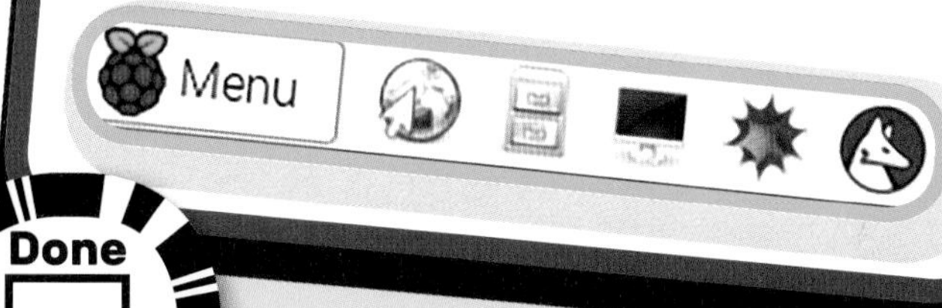

Done

3. Test the sound

1. Connect the speaker to the Raspberry Pi.

2. Create a new folder called whoopee with the following command...

```
mkdir whoopee
```

3. Enter the folder with...

```
cd whoopee
```

We're going to need a sample sound file for this project, so we'll download one.

4. Download the burp sample with the following command...

```
wget http://rpf.io/burp -O
burp.wave
```

This will download the sample into the whoopee folder and rename it to `burp.wav`

5. Now test that you can play the sound file using omxplayer by typing...

```
omxplayer burp.wav
```

You should hear it from the speakers or headphones connected to your Raspberry Pi.

If you can't hear anything, make sure that your speakers are connected correctly. If this still doesn't work, you'll need to change your audio configuration. To switch audio to the headphone jack, return to the Terminal window and type the following command...

```
amixer cset numid=3 1
```

Then try running the `omxplayer burp.wav` command again. Once the sound is working, it's time to write the program itself.

4. Write the program in Python

1. To write your Python program you will need to open the Python programming environment IDLE3 from the command line. To do this type the following command...

```
sudo idle3 &
```

2. Once IDLE3 has opened, click on File and New Window. This will open a blank file. Click on File and Save As and name the file whoopee.py.

3. Type in the following program precisely. (Pay special attention to indentation and lower-case/upper-case letters!)

```
import time
import RPi.GPIO as GPIO
import os

GPIO.setmode(GPIO.BCM)
GPIO.setup(2,GPIO.IN)
while True:
    if GPIO.input(2) == False:
        os.system("omxplayer burp.wav")
    time.sleep(0.5)
```

4. Save the file by clicking on File and Save.

5. Run the program by clicking on Run and Run Module (shortcut: F5)

6. Finally, test your program – check that the sample plays when the foil contacts are gently pushed together. If it's all working then it's ready to go!

5. Setting it up

- Carefully place your WhooPi Cushion where your victim will sit on it (obviously!), but not under a really heavy cushion where it will squash it straight away.
- The tricky bit is setting up the Pi so that it can't be seen – remember, you'll need a plug socket unless you are using a battery for your Pi.
- Hide it, run the program, and wait.

Done

Hint

Whistle tunelessly and look around at the ceiling. This will make you look innocent and attract potential victims.

Other things to try...

Take your Raspberry Pi tricks to the next level with these ideas!

- Use other noises/tunes/samples. They need to be in 'wav' format
- Record your own voice. How about recording your own voice shouting, "Help! You're sitting on me!"
- Record your parents when they're telling you off and put the WhooPi Cushion under their pillow in bed with the new sample. Note: This may lead to loss of pocket money or worse. Especially if you record it and put it on YouTube.

THIS RASPBERRY PI PROJECT IS SO MUCH FUN!

Visit...
www.raspberrypi.org/resources/make/

for more ideas and for help on using buttons and input devices on the Pi. This is also the first place to go if you are stuck or something's not working. You can also ask questions and get help on our forums at...

www.raspberrypi.org/forums

Anagram Acrostic

Unscramble the words!

Timed Challenge!

Can you work out these anagrams? You've got just **10 minutes** to test yourself! All of them are associated with Raspberry Pi or computing in general. As you fill the answers into the grid, the letters in the **shaded green column** may help you out...

- ENACT FIRM
- RENT MAIL
- TSK DOPE
- RAG ROMP
- DO BAKERY
- ERRS BOW
- SLEW RISE
- WOK TERN
- NOT HYP
- BRIAN SPA
- NIL UX

ZZZAP! I'VE SCRAMBLED ALL THE WORDS WITH MY RAYGUN!

Find the answers on page 78

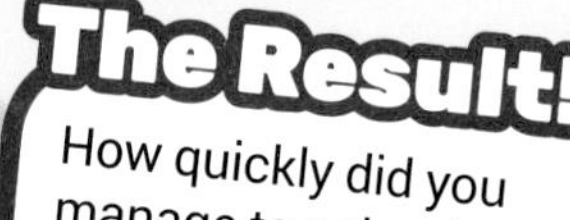

The Result!

How quickly did you manage to solve the puzzle? Write in your time below...

____

Python Puzzle

Mixed up words in Python code!

How good is your Python – and your codebreaking? Here's a challenge to find out. There are **8 Python functions** – each of them prints out a single word when run. Can you work out what all 8 words are, and what they have in common? If you get really stuck, you can use Python on your Pi to run them, but try to work it out without cheating first!

THE FIRST ONE IS REALLY EASY, TO GET YOU STARTED!

```
def test1 ():
    text1 = "apple"
    print(text1)

def test2 ():
    text1 = "eniregnat"
    text2 = ""
    for x in range (len (text1)):
        text2 += text1[-1 * x - 1]
    print(text2)

def test3 ():
    text1 = "oag"
    text2 = "ren"
    text3 = ""
    for x in range (len (text1)):
        text3 += text1[x]
        text3 += text2[-1 * x]
    print(text3)

def test4 ():
    text1 = "bceyabrlkr"
    text2 = ""
    x = 0
    for y in range (len (text1)):
        text2 += text1[x]
        x += 7
        x %= len (text1)
    print(text2)

def test5 ():
    list1 = [109, 97, 110, 103, 111]
    text1 = ""
    for x in range (len (list1)):
        text1 += chr (list1[x])
    print(text1)
```

```
def test6 ():
    text1 = "dsulfrw"
    text2 = ""
    for x in range (len (text1)):
        text2 += chr (ord (text1[x]) - 3)
    print(text2)

def test7 ():
    text1 = "($+!$7,+"
    text2 = ""
    for x in range (len (text1)):
        text2 += chr (ord (text1[x]) ^ 69)
    print(text2)

def test8 ():
    list1 = [0,39,16,10,0,40,18,19,18,24,0,4,
27,30,17,34,0,35,41,19,24,40,13,0,17]
    text1 = ""
    a = int (pow (len (list1), 0.5))
    for x in range (1 + max (list1) // a):
        z = sum (list1) // a
        for y in range (a):
            if y + x * a in list1:
                z += 1 << (a - 1 - y)
        text1 += chr (z)
    print(text1)
```

Learn to code with Python, visit...
www.raspberrypi.org/learning/python-intro

Puzzle Answers

Page 30
Spot the Difference – Pi

Did you manage to find the 12 differences in the board?

Page 31
Big Pi Wordsearch

How fast did you managed to find the 17 hidden words?

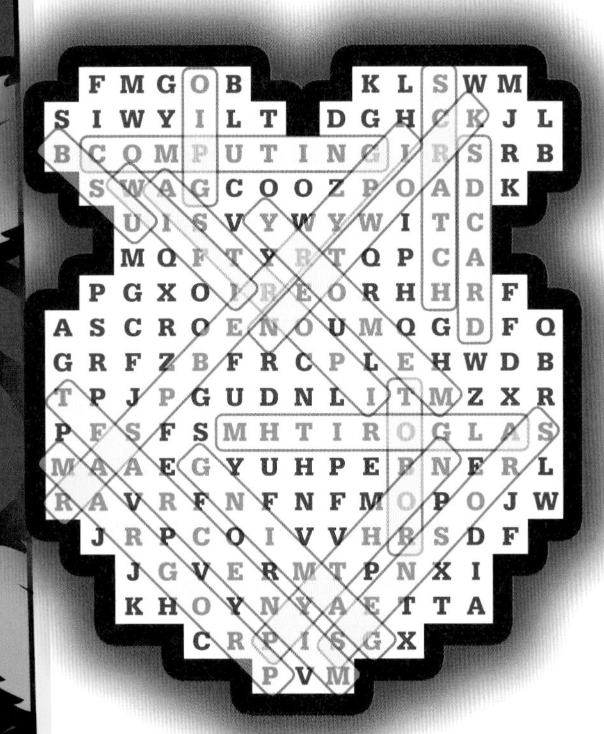

Page 40
Spot the Difference – Space

12 more tricky differences here...

Page 46
Space Maze

Here's where you should have gone...

Page 50
Wheel of Pi

The phrase was: CODE MINECRAFT WITH PI

Page 76
Anagram Acrostic

The answers were: Minecraft, Terminal, Desktop, Program, Keyboard, Browser, Wireless, Network, Python, Raspbian, Linux. The secret phrase was Raspberry Pi.

Page 77
Python Puzzle

Answers: apple, tangerine, orange, blackberry, mango, apricot, mandarin, raspberry.